Growth, Distribution and Innovations

As the political and economic landscape of capitalism has evolved through the years, growth theory has compensated for the changes with new concepts and tools of analysis. In this series of lectures, Amit Bhaduri calls for a more imaginative approach to understanding the process of capitalistic growth through recombining the insights of the great classical and modern economists.

In this concise and engaging book, Bhaduri sketches an alternative approach to mainstream growth theory, incorporating the role of division of labour, innovation and market structure according to Smith, Marx and Schumpeter, the role of class distribution of income in growth according to Ricardo, and the principle of effective demand according to Keynes and Kalecki. A formal framework of analysis which can accommodate these diverse insights is outlined.

Drawing on contemporary issues such as the role of competition policy, labour market flexibility and intellectual property rights regime in influencing the rate of economic growth, this volume will be ideal for advanced students of macroeconomics. It will also be of interest to anyone engaged with growth and distribution theory and technical innovation.

Amit Bhaduri is internationally selected professor in Pavia University, Italy and also visiting professor in the Council for Social Development, Delhi.

The Graz Schumpeter Lectures

Previous titles in the series:

1 **Evolutionary Economics and Creative Destruction**
J. Stanley Metcalfe

2 **Knowledge, Institutions and Evolution in Economics**
Brian J. Loasby

3 **Schumpeter and the Endogeneity of Technology**
Some American perspectives
Nathan Rosenberg

4 **Consumption Takes Time**
Implications for economic theory
Ian Steedman

5 **Exchange Rates and International Finance Markets**
An asset-theoretic approach with Schumpeterian perspective
Erich W. Streissler

6 **An Unholy Trinity**
Labor, capital and land in the new economy
Duncan K. Foley

7 **Politics and Economics in the History of the European Union**
Alan S. Milward

8 **The Dynamics of Industrial Capitalism**
Schumpeter, Chandler, and the new economy
Richard N. Langlois

9 **Growth, Distribution and Innovations**
Understanding their interrelations
Amit Bhaduri

For more information, please visit the Graz Schumpeter Society's website:
http://homepage.univie.ac.at/Bernd.Brandl/schumpeter/schumpeter.html

Growth, Distribution and Innovations

Understanding their interrelations

Amit Bhaduri

Routledge
Taylor & Francis Group
LONDON AND NEW YORK

First published 2007
by Routledge
2 Park Square, Milton Park, Abingdon, Oxon OX14 4RN

Simultaneously published in the USA and Canada
by Routledge
711 Third Avenue, New York, NY 10017, USA

Routledge is an imprint of the Taylor & Francis Group, an informa business

Typeset in Times by
HWA Text and Data Management, Tunbridge Wells

British Library Cataloguing in Publication Data
A catalogue record for this book is available from the British Library

Library of Congress Cataloging-in-Publication Data
Bhaduri, Amit.
Growth, distribution and innovations : understanding their interrelations / Amit Bhaduri
p. cm.
Includes bibliographical references and index.
1. Technological innovations–Economic aspects. 2. Economic development. 3. Diffusions of innovations. I. Title.
HC79.T4B495 2007
338′.064–dc22 2006033163

ISBN10: 0–415–42108–X (hbk)
ISBN10: 0–203–96287–7 (ebk)

ISBN13: 978–0–415–42108–9 (hbk)
ISBN13: 978–0–203–96287–9 (ebk)

Contents

Preface

This slim volume is a revised, and hopefully a more coherent version of the set of four lectures I had delivered at the University of Graz in the summer of 2005 at the invitation of the Schumpeter Society. Several members of the Economics Faculty as well as of the University administration went out of their way to make this a wonderfully pleasant and memorable stay. That pleasant memory lingers on, and my heartfelt thanks to all of them, especially to Professors Christian Gerke and Heinz Kurz, who presumably had meticulously planned every detail to make our stay in Graz so enjoyable.

There were lively discussions and critical observations with active participation of many from the audience, especially on the last day of the lectures. It helped to remove several cobwebs in my thinking, and they cannot be blamed for the remaining ones. I have tried to incorporate in this revised version some of the points that appear to me to be both valid and reasonably tractable within the bounds of these lectures. I am especially grateful to Nicholas Baigent, Christian Gerke, Christian Lager and Heinz Kurz for comments. I must especially thank Roland Wendner for saving me from an embarrassing error.

Finally, I should mention that these lectures try to bring together the problems of economic growth, distribution and innovations within a coherent macroeconomic framework. The purpose is to challenge received wisdom and mainstream orthodoxy on several fronts. This volume would have served its purpose if it can stimulate the reader to think outside the box of ruling economic orthodoxy.

Amit Bhaduri
New Delhi
23 September 2006

1 Lecture I: Introduction

Defining our approach to the problem of economic growth

It is a great honour for me to have been invited this year (2005) to deliver the lectures in memory of Josef Schumpeter. This distinguished series of lectures, which the Schumpeter Society organizes every year in Graz, is a testimony to the lasting intellectual legacy of Schumpeter. He is undoubtedly one of the most influential economists of our time. His powerful vision of capitalism, especially of capitalistic growth and fluctuations driven largely by innovations, has left a permanent mark on the history of economic thought. There is also no other place more suitable than Graz to honour the memory of this great economist. Graz was an important milestone in Schumpeter's intellectual journey as a professional economist. Nevertheless, the ideas he propagated soon went far beyond the confines of any particular place or time. They bear the mark of greatness, precisely because of their universalism; their lasting appeal goes beyond any particular theoretical model or set of arguments he might have used to elaborate his vision of capitalism. In the preface to the English edition of his *Theory of Economic Development* (Schumpeter, 1961; original in 1911 in German), which soon made him famous far beyond the German-speaking world, Schumpeter writes that the origin of many of his ideas of development goes back to 1907. Almost a century later, we are here today, celebrating the lasting power of those ideas.

The universalism of Schumpeter's vision of capitalism nevertheless implies something rather specific for me. Although Schumpeter himself often expressed a somewhat romantic view of capitalism in his various writings, it would be wrong to read his work through an ideologically coloured glass. To illustrate this point, I can do no better than to quote what Schumpeter wrote of Marx. 'No serious argument ever supports any "ism" unconditionally. To say that Marx stripped of phrases, admits of interpretation in a conservative sense is only saying that he can be taken seriously' (Schumpeter, 1943: 58). What Schumpeter had said of Marx applies equally well to him. His ideas relating to capitalistic growth and development defy narrow ideological labels of one kind or another. By the same criterion, Schumpeter too has to be taken seriously by every economist interested in the theory of capitalist development.

At the same time, nearly a hundred years later I remain somewhat puzzled as I try to formulate more precisely the core of these ideas in terms of modern economic analysis. His ideas are grand but elusive. Perhaps there is a reason for this. Schumpeter's particular ideas scattered through his various writings (1939, 1943, 1961), e.g. the circular flow characterizing economic life, the historic specificity of business fluctuations despite some observed patterns, innovations as the driving force of capitalistic growth, the consequences of competitive and oligopolistic market structures for innovations, the systemic competition between capitalism and socialism and many other related ideas, are all component parts of this grand vision, but they do not appear to be neatly arranged in a coherent analytical framework. At the same time, I recognize that this way of looking at Schumpeter's ideas might be misleading, because the general impression I form from his writings suggests that his uniqueness comes always from viewing these ideas or constructs as illuminating particular aspects of the dynamics of capitalism. They are theoretical constructs trying to place the evolving complex economic system called 'capitalism' on the coordinates of economic analysis. Schumpeter never lost sight of the fact that the capitalist system is constantly in disequilibrium, in the flux of history. Economics and history tend to merge with one another in his writings. Schumpeter's ideas appear elusive in places, because we have become accustomed to a somewhat different mode of theorizing. Standard economic analysis these days demands constructing idealized equilibrium situations as points of reference for analysing a system. Schumpeter did not disown this method, e.g. his use of the notion of 'circular flows'. He considered it necessary even when the economic system is constantly in disequilibrium. However he was also painfully aware of the tyranny of equilibrium analysis. This is apparent from his writings insofar as he recognized that this method often tends to make disequilibrium a somewhat inessential sideshow of the equilibrium analysis. It is this tension arising from using equilibrium analysis to understand a system constantly in disequilibrium that gives Schumpeter's writings their unique flavour. It also leaves them open to two almost diametrically opposite interpretations. If we lean on the equilibrium side, Schumpeter belongs to the orthodox neo-classical intellectual tradition in economics which highlights analytics and underplays history. It is a tradition which he himself valued greatly. On the other hand, if his insistence on disequilibrium economics placed in a historical flux is considered as the core of his vision, he belongs to quite a different tradition, far more difficult to reconcile with the current orthodoxy in economics. A hundred years later, Schumpeter remains an inspiring, and yet somewhat enigmatic economist precisely on this account. He tried to face the ultimate challenge to economic theorizing: the need to break out of the severe limits that equilibrium analysis imposed artificially on a historical process constantly in disequilibrium. Nowhere is this more apparent than in his justly famous metaphor of capitalism as a process of "creative destruction". He uses and discusses this concept under various names, and in different guises in many places. In particular, he devotes a short chapter to sum it up in one of his later works (Schumpeter, 1943: 81–6).

One particular aspect of the Schumpeterian process of creative destruction is particularly useful for illuminating the relation between equilibrium and disequilibrium analysis. It would also be relevant later from the point of view that I wish to adopt for analysing the problem of growth and innovation in these lectures. Schumpeter criticizes standard approaches in terms of equilibrium by pointing out that, "… the problem that is usually being visualised is how capitalism administers existing structures, whereas the relevant problem is how it creates and destroys them". And, he gives an interesting hint as to why this is so: "A system – any system economic or other – that at every given point of time fully utilises its possibilities to the best advantage may yet in the long run be inferior to a system that does so at no given point of time, because the latter's failure to do so may be a condition for the level or speed of long run performance" (Schumpeter, 1943: 83). Translated into modern jargon, this means that a series of successive efficient static equilibria might fail to characterize a dynamically efficient path.

The observation leaves room for at least three different interpretations. Since they all influence from different angles our approach to the problem of capitalistic growth, it would be worthwhile to make them explicit. According to the first interpretation, static allocative efficiency goes with an idealized form of market structure, namely perfect competition. By the 'fundamental theorem of welfare economics', perfect competition is celebrated as (Pareto-)optimal, statically the most efficient system. Yet Schumpeter recognized that a competitive system of innumerable atomistic firms with free entry and exit is not necessarily the best performing system over time, particularly in terms of its ability to promote innovations. Therefore, for him, the question remains open as to which type of the market structure is most effective under what circumstances for promoting long-run growth through generating and diffusing new technologies, new products, and organizational innovations.

Second, we could interpret Schumpeter's concern about the long-term performance of the economic system in terms of analytical categories like increasing returns or related concepts like cumulative causation, positive feedbacks etc. A system can deviate substantially from its static performance over time in the presence of increasing returns (Young, 1928). The origin of increasing returns is 'externalities' in some form which can arise from many different sources. For example, few would contest the idea central to many current models of endogenous growth that the spill-over of productive knowledge, which no regime of intellectual property rights succeeds in eliminating completely, is an important source of such increasing returns. In short, what appears to be a serious immediate private loss due to the leakage of knowledge to, say, business rivals, may still turn out to be a social gain that makes the system more efficient in the long run. Similarly, in the presence of increasing returns, a small initial advantage that accrues by chance to a statically less-efficient unit, say a firm, can continue to be magnified over time through a series of positive feedbacks. This might lead to the statically less-efficient system turning out to be dynamically more efficient (Arthur, 1994). As an illustrative story, consider the case of an innovation in which some small chance factor intervenes, say the principal scientist tracking this technology in the rival firm suddenly falls ill. This delays its diffusion or adoption by this rival firm, reducing the social benefit from diffusion. However, this chance factor also permits the innovating firm to reap monopoly profit for longer. Suppose now the innovating firm enjoying the additional profit due to the chance factor, invests it successfully in further improvement of the technology. That would make the technology path more efficient over time although it appeared less efficient initially due to slower diffusion of innovations. In other words, a system that appears statically inefficient at a certain point of time may increasingly gain efficiency through a small chance factor magnified by positive feedbacks. The general message of this example is clear. We need to recognize our intrinsic inability to predict the behaviour of a system subject to increasing returns. Marshall (1920: Appendix H) was well aware of these uncomfortable possibilities created by increasing returns, which drive a complete wedge between the static and the dynamic efficiency properties, making prediction based on current information particularly hazardous about the dynamic efficiency of a system which is subject to increasing returns.

A third interpretation is also plausible. Schumpeter looked optimistically at capitalism as being driven by a continuous process of creative destruction. According to him it is the source of tremendous dynamic efficiency. However, the reverse side of creative destruction is the wastage of resources it implies, at times very considerable and long lasting. Unless destruction is automatically matched by creation with negligible time-lag, creative destruction might be dominated by destruction, and little creation. For instance, in advanced market economies a significant section of the labour force often remains unemployed for considerable periods, while wide-spread acute poverty is the rule rather than the exception insofar as underdeveloped market economies are concerned. Schumpeter might have thought of unemployment simply as destructive aspects of the system, which would be dominated by the dynamic efficiency of creativity of the capitalist system. This might have been one of the reasons why Schumpeter was not persuaded by the Keynesian theory of effective demand, and tended to underplay its importance. Like many neo-classical economists he did not accept deficient effective demand as a chronic problem afflicting capitalism. For him, unemployment seemed to represent merely the destructive aspect of the creativity of the capitalist system, a view that would fit well with the notion of unemployment as essentially 'frictional'; a mere temporary lapse from the automatic full-employment position due to technological changes. It necessarily arises in the course of structural changes, obsolescence of skills, and substitution of old by new products or methods of production in any dynamic economy. However, how plausible one finds this view depends on whether one sees unemployment due to deficient demand as an intrinsic property of the system, or as a temporary lapse from the full employment equilibrium.

The rate of unemployment is known to increase and persist in fairly long spells, without any corresponding increase in the tempo of structural change. This renders the view of frictional unemployment less persuasive. Without destruction leading automatically to creation in these circumstances, we cannot ignore without further scrutiny the negative effects of the destruction in the form of persisting unemployment. In particular, when we recognize that it might have the effect of retarding subsequent creation or regeneration of industries. For instance the unemployed fraction of the labour force, often disproportionately large in the younger age groups, might lose out, insofar as they do not get the chance to learn from work experience. In consequence, the impetus for increasing returns relating to what these days are called 'human capital' formation gets weakened. This may also fail to raise the long run growth potential of the economy, through 'learning by doing', on-the-job skill acquisition, entry of the young to knowledge-based industries, and production of new goods and services. With high and lasting unemployment especially in the younger age groups, these negative effects resulting from a lowering of the general level of capability of the labour force may even persist long after the unemployment is removed (Blackburn, 1999). This persistence of the effect even after the removal of the cause is the typical phenomenon of 'hysteresis'. In several important instances, the negative impact of hysteresis may continue to weaken the subsequent creative process. It is arguable that in his admiration for the dynamism of capitalism, Schumpeter tended to underplay these negative aspects. In an even more brutal form, the negative effect of hysteresis appears in many low-income market economies. The poorest sections of the population become economically marginalized mostly through their lack of access to productive assets, education and acquisition of skills through employment in a market economy. The Schumpeterian process of destruction operates by denying them access to reasonable livelihood, training and skill through the usual mechanism of price rationing in the market economy. There is little doubt that this marginalization weakens considerably the impetus for the dynamic, creative aspect of capitalism, leaving an underdeveloped market economy trapped in poverty. In this case it would be hard to agree with Schumpeter that regenerative creation follows automatically the destruction that takes place through marginalization of a significant section of the population through the market mechanism.

On balance, we should recognize that creative destruction is a metaphor reminding us about the vitality of capitalism. And yet, it provides little more than powerful rhetoric about the process of capitalistic growth. For making use of it as a serious argument, we must take into account its two-sided nature. Like the double-edged sword that can wound the very person who wields it, creative destruction can help or hinder the performance of the market system with the persistence or hysteresis effect of destruction dominating at times the positive effects of creation. In this way we need to apply Schumpeter's dictum about Marx to his own theory. No serious argument, including the analytical content of 'creative destruction', supports any particular 'ism' including capitalism and the market mechanism. Indeed, it is the two-sided nature of market competition that we will try to capture analytically in a model presented later (Lecture III).

In Schumpeter's view competition among rival firms is the force driving the process of creative destruction. The number of competing firms would vary greatly, depending on the particular market structure. For example, they would be numerous under nearly perfect competition, but few under oligopoly. Since innovation and entrepreneurship are the two most important components of creative destruction identified by Schumpeter, their interaction with the market structure occupies in his scheme a central position for understanding the dynamism of capitalism. He was rightly convinced that understanding the efficiency of the market mechanism from this angle was far more important than worrying about its static allocative efficiency. In this respect he was distinctly different from most modern neo-classical economists who are preoccupied with the static allocative efficiency of the price mechanism. One might even say Schumpeter was neo-classical in his neglect of the problem of effective demand, but anti-neo-classical in his underplaying the problem of the allocative efficiency of the price mechanism.

In standard theory, the virtues of competitive equilibrium stem from its Pareto-optimal properties. In his early writings, Schumpeter (1961) took the classical view that a nearly competitive market structure, characterized by many firms with free entry and exit, has the particular virtue of turning technological competition into price competition through the diffusion of technology. The first-mover individual entrepreneur adopts a new innovation for cutting costs, and forces other firms sooner rather than later to adopt the same technology for survival under the discipline of competition. Therefore, the 'abnormal' profit from each particular round of innovation, being related to the speed of diffusion of technology, is transient. Implicit in this scheme is the idea that the generation of productive knowledge and innovation are largely independent, individual activities. It is transformed into commercially viable technology by a combination of productive knowledge with entrepreneurship. In this scheme, the generation of innovation is viewed mostly as an exogenous and somewhat sporadic process propelled by individual initiative, whereas its adoption and diffusion are treated as endogenous processes governed by the system of competitive capitalism of many small firms.

Fortunately, Schumpeter did not suffer from the typical narrowness that often makes academic economics precise but largely irrelevant. He was acutely aware of the changing character of capitalism, and its evolution through changing market forms. Even during his lifetime, the dominating presence of large corporations was becoming increasingly apparent. It was proceeding through various organizational innovations that were changing property relations through extensive ownership of stocks and limited liability that separated management from ownership. With a considerable degree of prescience, Schumpeter discussed its impact on the generation of productive knowledge in the society. In his later writings, with rare insight, Schumpeter (1943) pointed out how routine in-house research by large oligopolistic corporations influence the very way technology is generated on the one hand, and gets diffused on the other. His overall verdict seems to have been that the pressure of competition is no less acute in a market with a few large corporations than in the case of a market with many small firms. It usually forces rival corporations for sheer survival to join the technological race through routine research. However, this race for generating new technologies through routine research has its own peculiar characteristics which are quite different from those under near-perfect competition. The development of technology through routine research typically tends to be incremental and at times even trivial in content. This process of minor differentiation in technology can be seen as the counterpart of monopolistic product differentiation. The incremental improvements through routine research are directed primarily towards solving specific problems faced on a day-to-day basis by a corporation. It is meant to serve the specific interests of a particular business, rather than being driven by scientific curiosity. In this way, not only the diffusion but also the generation of innovation under corporate capitalism tends to become almost like any other routine economic activity related to commodity production. It is undertaken usually in the research and development department of a firm by a group of specialists with modest targets set at regular intervals to protect and further the profits of the corporation.

From this perspective, the generation of new technologies and the endogeneity of technical progress through routine research become standardized processes, almost like the production of any other commodity. This is largely a consequence of the corporate nature of capitalism. Analytically, this would seem to justify to an extent the prevalent practice of using 'production functions' for the generation of new productive knowledge on par with any other commodity in many contemporary neo-classical growth models (e.g. Frankel, 1962; Lucas, 1988; Romer, 1986, 1990). However, this formally convenient representation of the production of knowledge generated through routine research should be recognized as the outcome of the corporate structure of capitalism. This was Schumpeter's fundamental insight. Therefore, using the representation of knowledge production on par with commodity production might suit better the market structure of oligopoly than that of perfect competition.

In any case, capturing formally the production of knowledge in a plausible manner is an extremely difficult task. Unlike the production of normal commodities, it is complicated by the uncertainty that surrounds the outcome of even routine research. We would take the position in these lectures that the diffusion of technology in contrast to its generation is a more tractable economic process. Irrespective of whether the threat of competition comes from a few or from many, the process of diffusion would be driven by competition among rival capitalist firms. The differences in market structure would make a difference to this process insofar as they shape differently the link that is forged between technological competition on the one hand, and price competition on the other. Under nearly perfect competition, as the innovating firm lowers its price in line with its lower unit (or marginal) cost of production due to the improved technology, numerous other small firms follow as price-takers for survival. In this way, prices are lowered due to lower production cost. However, under oligopoly, firms as price-makers can behave more independently to weaken this link between technological and price competition. For instance, an innovating oligopolistic firm might behave in a manner similar to the semi-competitive firm, and cut price in line with lower production cost to obtain a greater market share. Alternatively, it might enjoy a higher profit margin at the lower production cost, insofar as the oligopolistic market structure provides it with more space of not engaging in direct price competition with rival firms. It might choose for instance to protect its higher profit by trying to maintain its technological lead in one way or another. Therefore, in markets with competitive characteristics, the benefits of innovation are passed largely on to the consumers through lower prices. However in the latter case of pronounced oligopolistic or monopolistic characteristics, the benefits of technological innovation might be retained by the innovating firm through higher profit margin. This is not to deny that the technology would get diffused through spill-over under imperfect property right regimes, or get overtaken by a superior technology, but the oligopolistic firm has the option of not engaging directly in price competition.

We could therefore focus fruitfully on the nature of this link between price competition and technological competition under different market structures. At least for examining the impact of technology diffusion on growth, this seems to be a more helpful approach than the conventional distinction between perfect and imperfect competition. As a matter of fact, one could helpfully draw a simple distinction between nearly competitive market structures in which technological competition is more or less automatically converted into price competition and various monopolistic forms of the market structures, in which the benefits of technological progress are retained as long as possible by the innovating firm in the form of higher profit margin. This operational distinction between 'competitive' and 'mature' market forms of momopoly capitalism (Steindl, 1952) would serve to capture in a parsimonious way the inter-linkage between technical progress and market structure.

The view that technological progress, and particularly its diffusion, is driven largely by intra-class rivalry among capitalist firms has a long shared intellectual history. All the great masters of classical political economy from Smith to Marx shared this view in one way or another. More than a century before Adam Smith (1776), William Petty (1662) had identified the process of increasing technical and social division of labour as the main force contributing to the wealth of nations. Both Petty and Smith recognized that the two aspects of the division of labour, its technical or micro aspect and its social or macro aspect, interact continuously to raise the level of productivity of labour. Increasing technical division of labour in Smith's celebrated example of the pin factory leads naturally to the idea of increasing returns at the micro-level of an enterprise, both through the scale effect, and over time, as each successive round of production leads to further learning by doing in various forms. Petty had given a similar example of making watches rather than pins. He pointed out how the technical division of labour internal to a firm combines with spatial agglomeration of the makers of the various components of watches to raise the productivity of labour for lowering costs, and foreshadowed the idea of Marshall's industrial district. While these are examples of the division of labour internal and external to the enterprise, the Physiocrats as well as the classical economists were also aware of the broader macro aspect of the social division of labour, i.e. how the labour force is distributed between agricultural and non-agricultural activities, and the role that occupational structure plays in raising the level of labour productivity through division of labour in the economy.

Division of labour entails specialization in production, and specialization involves exchange. Therefore, increasing division of labour requires the scope for exchange to be expanded continuously through the market. Smith's celebrated observation that the division of labour is limited by the extent of the market is quite natural from this point of view. In this way, by defining the scope for exchange possibilities he assigned to the size of the market the role of setting the limit to labour productivity through the extent of division of labour. This celebrated observation of Smith has a deeper and more fruitful interpretation from the modern perspective. The process of division of labour through various positive feedbacks leads to increasing returns and continuously decreasing costs over time from the supply side. Because this process of decreasing cost by itself has no natural limit, its limit has to be set from the demand side, i.e. the level of production that the purchasing power or the size of the market would be ready to absorb. This contrasts sharply with the neo-classical view in growth theory, which finds no place for aggregate demand or the size of the market (Solow, 1956; Swan, 1956). Some of the more recent neo-classical models try to incorporate the problem of increasing returns exclusively from the supply side, by balancing the diminishing returns to 'physical capital' against increasing returns to 'intellectual or human capital' (e.g. Barro and Sala-i-Martin, 1995).

In this respect a comparison with Ricardo's theory of growth (1817) might be instructive, because Ricardo looked at the problem from almost an opposite angle, focusing on increasing costs due to diminishing returns (Pasinetti, 1960). As the margin of cultivation is extended, land of lower fertility is brought into cultivation, making land subject to diminishing returns at the extensive margin. Rent being determined by the difference between the average and the marginal productivity of land, as the margin of cultivation increases, the share of rent in national income increases at the expense of profit due to diminishing returns on land. By making further the assumption, which might have been sociologically plausible for his time, that all profits of the capitalists but no rents of the landlords or wages of the workers are saved, Ricardo could link directly this process of redistribution of income from profit to rent with saving contributing to the wage fund. He argued that the gradual drying up of profits as the source of investible fund, would limit growth to drive the economy towards its ultimate stationary state. Land as a symbol of all primary factors not produced within the system sets the ultimate limit to growth. However, this Ricardian dynamics leading to the classical stationary state requires strong implicit assumptions (cf. Bhaduri and Harris, 1987). In particular, Ricardo was assuming that all the saving of the economy in the form of additional wage fund would be automatically invested back in each period for extending the margin of cultivation until profit dwindles to zero. The consequence of this assumption that profit as saving is automatically invested was severe as it amounted to ignoring altogether the problem of demand. And, the observation of Smith that the size of the market sets the limit to the division of labour fell by the wayside.

Subsequent development in the theory of effective demand due to Keynes and Kalecki hinges precisely on the distinction between the decisions to save and to invest. We can see on hindsight that the Ricardian assumption of saving being automatically invested, leads to ignoring the problem of effective demand (Robinson, 1964). Since it leaves the size of the market with no role, the result is an exclusive supply-side view of growth, in which increasing marginal cost operates as the constraint on output. It is seldom realized that we have these two contrasting paradigms not merely in growth theory but almost in all branches of macro-economic theory. According to one perspective, output is limited by aggregate demand even if costs are decreasing, while the other view claims that output is limited by rising marginal cost. The distinction between Keynesian and classical unemployment also hinges essentially on this distinction (Malinvaud, 1977), and it breaks down if effective demand is endogenized into the system (Bhaduri, 1983).

Unfortunately, ignoring altogether the problem of effective demand has become the rule rather than the exception in mainstream neo-classical growth theory. Through the very assumption made by Ricardo that all saving is automatically invested, starting with Solow (1956) and Swan (1956), the neo-classical tradition in growth theory construct long-term growth models in which diminishing returns, i.e. rising marginal cost in one form or another, appears as the limiting factor. In this case, economic growth would ultimately be constrained from the supply side by the relative scarcity of some natural or primary resource like land or labour. From this point of view the central result of post-war neo-classical growth theory could be considered a reinvention of the Ricardian result. In models following this tradition, 'capital' is considered as a factor of production in an aggregate production function on par with labour. However, being a produced means of production, capital becomes augmentable through saving, as it automatically gets invested. In place of Ricardian land, labour is the only primary, nonproduced factor in the system with the difference that labour grows at some exogenous rate. The model then shows that the economy converges to the exogenously given growth rate of labour due to diminishing returns to capital operating on the intensive margin. Thus, when capital grows faster (slower) than labour, a rising ratio of capital to labour through substitution between them leads to diminishing return to capital, and falling (rising) marginal product of capital until this ratio stabilizes with both capital and labour growing at the same rate. While the Ricardian economy converges to a zero growth rate in the stationary state due of the given amount of land, the Solow–Swan economy converges to a steady state determined by the exogenously given growth rate of labour.[1]

In different ways, both the Keynesian and the neo-classical models of growth owe something to Ricardo. The Ricardian idea of inter-class distribution between profit and rent, recast in a more contemporary context as the problem of distribution between wage and profit, features as an important variable influencing the saving rate in post-Keynesian growth theory. Thus, the link between functional or class distribution of income affecting saving and growth, which Ricardo had emphasized reappears prominently in many post-Keynesian growth models (e.g. Kahn, 1959; Kaldor, 1957; Pasinetti, 1962; Robinson, 1956, 1962). And yet, these models are not at all in the Ricardian tradition, insofar as they deal with the problem of aggregate demand by distinguishing investment from saving. On the other hand, the defining characteristics of neo-classical growth models are almost precisely the opposite. They ignore altogether the problem of aggregate demand, and usually prefer not to deal with the problem of class distribution. Instead, they focus exclusively on the supply side characterized by diminishing returns to the factors of production, and rising marginal costs. However, from the Ricardian perspective this neo-classical construction is misleadingly over-simplistic. It ignores all the complications of value theory that Ricardo had to face, and oversimplifies misleadingly the concept of 'capital' as a factor of production. Based on this logically indefensible concept of capital, it distorts the Ricardian theory of rent into a generalized marginal productivity theory of distribution. This requires assuming diminishing returns to all factors of production operating on the intensive margin through factor substitution, rather than extensive margin through fixed proportion with varying land quality.

It should be well known by now that the result derived within the neo-classical analytical framework of an aggregate production function in labour and capital is logically insecure, because it cannot be extended beyond a one-commodity world. In his search for an 'invariant measure of value', Ricardo had already become aware of the problem that no measure of 'capital' is possible independent of the class distribution of income.

Beyond the one-commodity world, the capital theoretic problems that beset the aggregate production function render the central mechanism of factor substitution and the associated marginal productivity theory of distribution logically unacceptable. This has been dramatized in the controversies in capital theory by the problem of 'reswitching of techniques'. It showed conclusively the possibility that same technique might be most profitable at two widely different wage rates, and yet, for the intermediate range of wage rates another technique might be more profitable. This demolished the notion that there is a monotonic inverse relation between the rate of profit and the 'capital intensity' of techniques. As a result, neither the marginal productivity theory of distribution nor the general claim that the rate of profit is an index of the relative scarcity of capital is logically sustainable outside a one-commodity world (Sraffa, 1960; Samuelson, 1966; Pasinetti, 2000).

In poetic language, it should have been 'after such knowledge, what forgiveness (for the aggregate production function)?' And yet, the convenience of formal manipulation seems to take precedence over logical substance, and the aggregate production function, perhaps because of its easily manipulated properties, continues to dominate mainstream models of growth. The depiction of the supply side through an aggregate production function restricts it to a one-commodity model, while the Keynesian problem of effective demand is ruled out through the assumption of a single agent for whom saving and investment are one and the same decision. In effect, therefore we have a misleadingly reductionist one-agent, one-commodity model for analysing the problem of modern capitalistic growth! And, it is within this over-simplified framework that the Ricardian answer to the question of what limits growth has been reinvented as a central result in post-war neo-classical growth theory.

The classical tradition for understanding growth is by no means confined to Ricardo. Smith had considered it from a different perspective, and in several ways Marx changed radically the framework for analysing economic growth. With his emphasis on the importance of historical categories in economic analysis, Marx (1867) tried to isolate the particular features that characterize capitalistic growth. Since he started with commodity production for the market as the most basic feature of capitalism, he naturally encountered the problem of whether an adequately large size of the market would exist for absorbing the surplus product or 'surplus value' generated through the exploitation of the workers. In this respect, Smith was his predecessor in identifying the size of the market as a critical factor in limiting the extent of division of labour. Kalecki (1971) and Keynes (1936) were his successors in formulating precisely how the size of the market is determined through the level of investment.

Marx had tried to deal simultaneously with two related sets of issues. Micro-economically he confronted the problem of how the surplus is generated through the exploitation of individual workers by extending the hours of work, keeping the wage low etc, while its macro-economic counterpart was the realization of the total surplus generated by these means into monetary profits. This required selling the total surplus in a market of sufficiently large size. It follows from his analysis that a higher rate of surplus per labour produced by an increase in labour productivity through division of labour or innovation, without a corresponding increase in the real wage rate would raise surplus per worker, but might create simultaneously the problem of insufficient demand due to the limited size of the market. This is the problem of innovation viewed from the under-consumptionist perspective.

Marx's analytical construction remained opaque in this particular respect. The micro-aspect of exploitation resulting in the generation of surplus per individual worker and the macro-aspect of realization of aggregate surplus into profit needed to be connected by an analytical link. Marx's theory remained incomplete without specifying how the level of employment gets determined, because total surplus depends on the surplus generated per worker multiplied by the number of workers employed. The theory of effective demand of Kalecki and Keynes provided this missing link.

This line of reasoning has wider methodological significance. It points out how various microeconomic arguments about innovation often run into macro-economic fallacies of composition. For instance, an innovation causing a disproportionate increase in labour productivity in relation to the real wage rate might reduce unit production cost resulting in higher profit margin per unit of sale. And yet, it would not necessarily mean more total profits for all firms due to the possible barrier of insufficient, aggregate demand created by the redistribution of income in favour of profit through this innovation and a reduction in the overall size of the market.

Similarly, a hiatus might be created between an individual unit's decision to save more, or a firm's decision to cut wages, and its macro-economic impact through aggregate demand. These are all familiar Keynesian themes of the 'paradox of thrift', or the 'wage-cut controversy'. It also has a lesson particularly relevant for contemporary corporate capitalism. Higher labour productivity through downsizing the labour force, when carried out by a single corporation, might increase its market share and profit through lower unit cost. Nevertheless, when carried out by many corporations, it would reduce aggregate demand, possibly leading to lower total profit for business as a whole.

The presence of such fallacies of composition defines the very border separating micro- from macro-economic reasoning. And yet, it is overlooked systematically in contemporary macroeconomic models of growth in the neo-classical tradition. For instance, a class of models finds it good enough to proceed on the assumption of an all-seeing optimizing agent. His or her inter-temporally optimal saving plan is presumed to be invested automatically in each period, thereby avoiding all problems of deficiency in aggregate demand in this single-agent economy (e.g. Romer, 1996 for an exposition of the class of models). The literature on optimum saving was developed originally in the context of normative planning theory (Ramsey, 1928; Koopmans, 1965; Cass, 1965), and understandably abstracted from all problems of effective demand by assuming that saving is automatically invested in an idealized centrally planned economy. To transplant this model for understanding the working of a capitalist economy is thoroughly misleading. Another class of models of overlapping generations, despite their apparently more plausible assumptions regarding saving, misses again the same crucial point (Samuelson, 1958; Diamond, 1965). They fail to take into account the fact that the saving plans of households of various generations cannot be realized over time, without matching investment decisions in each period by the firms. In models of both optimum savings and of overlapping generations, the saving plans of households are always assumed to be realized through an exactly matching amount of investment by the firms, ruling out all problems of deficient demand. In short, the neo-classical mode of theorizing in all its modern variations keeps returning to some version of Say's law through its neglect of all problems of aggregate demand, and become purely supply side growth models despite a show of formal elegance. They are at best stories about what the economic growth could have been, if the capitalist economy had never suffered from unemployment and underutilization of capacities. Simplification, even radical simplification, must be allowed in theoretical formulations. However, they become misleading oversimplifications if they miss some of the defining characteristics of the problem under consideration. To discuss capitalistic growth in this framework devoid of all problems of aggregate demand, and a logically indefensible representation of the supply side through an aggregate production function, seem utterly hopeless for this task.

Therefore we would approach the problem of economic growth rejecting three misleading oversimplifications typical of mainstream growth theory. First, we have to depart from the neo-classical tradition of macroeconomic growth models, insofar as effective demand should be brought back to play a central role in the analysis. Second, we need to reject modelling the supply side through an aggregate production function, because it is logically misleading outside a one-commodity world. It also implies that we would not rely on the marginal productivity theory of distribution. Finally, we would incorporate into the analysis two intertwined aspects of competition that drive the pace of technical progress in a capitalist economy. It is a broader view of competition involving intra-class competition and inter-class competition. We would consider inter-class competition or conflict over the share of income which exerts an influence on productivity growth. Marx in particular had emphasized how labour-saving innovations create continuously a 'reserve army of labour' to keep real wage rate in check. This idea will be modified drastically to bring it in line with some stylized facts. Our emphasis will be on the share rather than the rate of wage, because it is not the wage rate but the wage share that has tended to be relatively stable in the course of capitalistic growth. Intra-class rivalry among capitalist firms under different market structures is intertwined with this inter-class conflict in driving both the generation and the diffusion innovations. Our formulation will focus on the more tractable process of diffusion rather than the generation of innovations. However, there is a paradox seldom noticed in linking technological diffusion with intra-class competition under different market structures and property rights regimes. The only water-tight case in which the private innovator retains all the benefits of his innovation is monopoly, simply because there are no rivals to whom the benefit can spill over through diffusion. And yet, such a monopolist would hardly have any incentive to innovate. (Some say this was a reason why the quality of consumers' goods remained so poor in former socialist countries.) On the other hand, competition among rivals that drives innovation will necessarily lead to some spill-over of technology, no matter how severe is the property rights regime. Since this would tend to discourage private investment in R and D, with the level of private investment lower than its socially optimal level (Arrow, 1962), an unavoidable paradox emerges. Intra-class competition is the life-force of innovative capitalism, and yet it also acts as a barrier to the pace of innovation! Through the modelling of intertwined intra- and inter-class conflict, our formulation will try to look at this paradoxical aspect of capitalistic growth from another angle.

2 Lecture II: Economic growth and the class distribution of income

In macro-economics the neoclassical mode of theorizing is generally most comfortable with the assumption of full employment of resources. The regular recurrence of unemployment on a significantly large scale stares awkwardly in the face of neo-classical economics as an uncomfortable fact. The coexistence of unemployment with a positive wage rate implying non-zero 'opportunity cost' for labour as an underutilized resource becomes hard to explain, without taking recourse to some particular explanations for the failure of appropriate signalling by the price mechanism. All neo-classical discussions of the unemployment problem are variations on this theme, except those fringe ideologues for whom the market or the price mechanism never fails anyway.

Against this background it is hardly surprising that continuous clearing of the labour market with full employment is assumed as a matter of routine in neo-classical growth models. Two rather different types of justification are usually offered in its defence. The first line of defence is sort of half-Keynesian. It pretends to recognize the Keynesian problem of effective demand, only to dodge it by assuming that the fiscal and monetary policies of a state that never fails would always maintain full employment. With the government being so perfect in insulating the economy from all the vagaries of the market, this assumption can hardly be considered an appropriate basis for understanding the process of growth under free-market capitalism.

The second line in defence of the full employment postulate makes a distinction between the short and the long run. It claims that unemployment is necessarily a short-run, transient problem. Phenomena like unemployment, inflation etc are said to arise because the price mechanism operates imperfectly in the short run on account of various rigidities, and imperfect information which get corrected over time. Therefore full employment is considered the natural assumption to make for the long-run growth theory, presumably because the price mechanism can be expected to have induced all the appropriate adjustments for ensuring the full employment of all resources in the long run. It is argued from this point of view that it would be legitimate to separate analytically the long-run problem of growth from short-term problems of fluctuations and unemployment.

Apart from the practical relevance or tenability of the built-in assumption of full employment equilibrium in the long run, the above mode of theorizing is questionable even on its own grounds. At the abstract theoretical level, there is nothing in the general equilibrium theory concerned with the properties of the price mechanism which either guarantees stability or specifies the length of time that would be taken by the price mechanism to reach equilibrium. Even if a unique and stable equilibrium is assumed to exist in a perfectly competitive economy under a host of unrealistic assumptions, the length of the adjustment time to equilibrium remains unspecified in theory. And, since theory cannot tell us how long is the long run, we all might well be dead before that long-run equilibrium is reached.

An even more serious problem arises when the concept of a long run is applied to growth theory, irrespective of what happens in the short run. An essential characteristic of a historical process is that the short-term events tend to leave their marks in various ways in shaping the long-run trend. It seems counter-productive to define an abstract potential output path characterized throughout by the full employment of resources, irrespective of what happens in several successive short periods. Two reasons for this can be identified easily. First, as mentioned in the last lecture, phenomena like 'hysteresis' arise when unemployment persists over several successive short runs. They leave their mark on the long-run potential growth path of output in various ways, e.g. through lower skill or 'human capital' formation, deskilling of the workers who have been unemployed for a length of time, failure to create sufficient new capacity etc. These effects usually persist long after the short-run cause ceases to operate. An even more obvious case of this phenomenon of hysteresis is the free entry and exit of firms. For instance, a policy of opening up speedily to international competition might make several domestic firms uncompetitive, and lead to their exit within a relatively short period. It might also alter the market structure, say from the 'monopolistic' competition of many firms to competition among a few under oligopoly. In this manner the particular path traced by successive short periods emerges as a generic feature of the system characterized by 'path dependence', which impacts in turn on the notion of long-run equilibrium of the system (cf. Bhaduri, 2002).

In insisting on this distinction between the short and the long run, it is seldom recognized that the long run gets invariably linked to the short run through the effect of the latter on expectation formation. For instance, an economy in recession for a few years would generally depress business expectations and investment. New capacity creation would slacken, and this would lower in turn the path of long-run potential output, linking inextricably the short with the long run.

Therefore, from the point of view I would adopt to study growth, there is no justification to isolate artificially the short from the long run. It cannot be maintained that inadequate effective demand is a short-run problem which can be isolated meaningfully from the long-run supply side of potential output path. In particular, without any prior assumption of continuous full employment the out of equilibrium adjustment dynamics needs to be studied first from the demand side. The supply side, which is influenced continuously by adjustments on the demand side, can be dealt with through an analysis of the behaviour of labour productivity in the course of these out-of-equilibrium demand adjustments.

The macro-dynamic adjustments on the demand side are assumed to be set in motion by the discrepancy between investment by the firms and saving by the households. This provides the usual Keynesian background to modelling the demand side. The influence of demand on the level of economic activity begins from the relatively non-controversial proposition that excess demand in the commodity market or excess of investment over saving tends usually to raise the quantity supplied or the price level or both. However, the speeds of adjustment of price and of quantity would in general be different. The cost-determined fix-price and the demand-determined flex-price models provide the two logical extremes. In the former the speed of price adjustment, and in the latter that of quantity adjustment, tends to be negligible (Hicks, 1965; Kalecki, 1971; Taylor, 1983). Indeed, this very difference in the speeds of adjustment was made use of by Marshall (1920) to distinguish between the 'short' and the 'long' period. He assumed that prices adjust faster than quantities, and therefore, adjustment of prices dominates over that of quantities, in the short period. A major novelty of the Keynesian short-period analysis was to reverse this order in the speeds of adjustment. Keynes argued that adjustment of quantity can be faster than that of price in situations of serious unemployment and excess capacity (Leijonhufvud, 1968).

Depending on whether we concentrate on quantity or on price adjustment, two somewhat different perspectives on the link between economic growth and distribution emerge within the broader Keynesian framework. Excess demand for commodities in a closed economy represented by an excess of investment over saving would normally lead to some quantity adjustment through higher capacity utilization. At the same time, in response to that excess demand some adjustment in the price level in relation to the money wage rate might also result, leading to a change in the real wage rate, and the distribution of income between the classes. Thus, the same Keynesian multiplier mechanism could operate in a more general framework to influence simultaneously both the level of output, and its distribution.

However, the redistribution of income between profits and wages through an adjustment in the real wage might affect in turn aggregate demand through two different channels. So long as the propensity to consume out-of-wage income is higher than that of out-of-profit income, a lowering of the real wage rate would tend to depress total consumption expenditure by redistributing income against the wage earners with a higher propensity to consume. At the same time an opposite effect might operate through the investment channel. Investment expenditure might get stimulated due to the lower the real wage, as it raises the profit margin per unit of sale. Depending on which of these effects dominates quantitatively in a closed economy, two alternative regimes of demand-led expansion emerge as distinct possibilities. The former would be led by greater consumption expenditure due to higher real wages, and the latter by greater investment expenditure stimulated by higher profitability. The former can be described as a wage-led, and the latter as a profit-led regime. Although for expositional simplicity we will restrict the subsequent analysis to a closed economy, note in passing that the emergence of the profit-led regime becomes more likely in an open economy, insofar as a lowering of the real wage rate helps to raise the level of aggregate demand by raising exports through greater international cost competitiveness.

The essential formalism of the two regimes is captured easily by normalizing the relevant variables with respect to full capacity output (Y^*) (Bhaduri and Marglin, 1990). Thus, under the classical assumption that no wage and a constant fraction ($1 > s > 0$) of profit is saved, the(normalized) saving of the economy is written as,

$$S = s.h.z \ , 1 \geq z, h \geq 0, \text{ where} \tag{1}$$

$h = \left(P/Y\right)$ = share of profit in output; and $z = \left(Y/Y^*\right)$ = degree of capacity utilization, with $Y^* = 1$, i.e. the normalized level of full capacity output.

For expositional simplicity we assume that (normalized) investment (I) depends positively on the same two variables, namely capacity utilization (z) and profit share (h), which by definition is related positively to profit margin (m) as,[2]

$$h = m/(1 + m) \ . \tag{2}$$

We make the assumption that expectations are formed in a climate of 'business as usual', so that 'conventions' rule expectations (Keynes, 1937). In this situation, expectations are formed mostly by extrapolating into the future the current state of affairs with few revisions. This corresponds formally to a state of quasi-static expectations. On this assumption the investment function is written as,

$$I = I(z,h), \quad I_z > 0, \quad I_h = (1+m)^2 I_m > 0. \tag{3}$$

By total differentiation of (1) and (3), the slope of the locus of saving–investment equality, or the IS curve, is derived in the *h-z* space as,

$$dz/dh = (I_h - sz)/(sh - I_z). \tag{4}$$

A positive slope of the IS curve in (4) means that a higher profit share (h) is associated with higher capacity utilization (z), characterizing a regime of profit-led expansion. Contrariwise, a negative slope of (4), associating a higher wage share (i.e. a lower profit share) with higher capacity utilization, corresponds to the wage-led path of expansion.

As is well known, the stability condition of the one-variable Keynesian income adjustment process, i.e. the convergence of the multiplier mechanism, is satisfied provided saving is more responsive than investment to changes in income. In that case, the denominator on the right-hand side of expression (4) is positive,

$$(sh - I_z) > 0. \tag{5}$$

Given that the denominator is positive by (5), a positive numerator of (4) requires,

$$(I_h - sh) > 0, \tag{6}$$

and yields a positive slope to the IS curve. Economically (6) implies that the stimulating effect of a higher profit share on investment outweighs its depressing effect on consumption expenditure, and this places the economy is on a profit-led path. If inequality (6) is reversed,

$$(I_h - sh) < 0, \tag{7}$$

and, for exactly the opposite reason, the economy is placed on a wage-led path.

Since saving equals investment on the IS curve, the commodity market is in equilibrium, and neither the degree of capacity utilization nor the distribution of income has a reason to change through the multiplier mechanism. Therefore, along the equilibrium *IS* curve, expansion of capacity is feasible only by treating the distribution of income (h) as an exogenous policy variable. Formally, this would reduce the underlying dynamical system to the usual single variable income determination process of the Keynesian theory. In this case, capacity utilization (z) adjusts to excess demand in the product market with income distribution (h) exogenously given. Using equations (1) and (3), this is represented by the quantity-adjustment equation,

$$dz/dt = \alpha\left[I\left(h,z\right) - shz\right], \quad \alpha > 0 \tag{8}$$

Where α is some arbitrary positive speed of adjustment, and h is exogenously given as a policy parameter.

However, the adjustment process in (8) could be reversed by taking recourse to the earlier Keynesian idea on the working of the multiplier mechanism through 'profit inflation' (Keynes, 1930). In this view, the distribution of income is determined endogenously through the interaction between the price level and the money wage rate in response to excess demand in the product market. This results in variation in the real wage rate, and a corresponding change in profit share. Although the price–money wage dynamics would normally be influenced by the state of demand in the product as well as in the labour market, the influence of the labour market is kept out of this argument by assumption, only to establish a parsimonious link between the price level and the money wage exclusively through the product market. Thus, when the price level and the money wage rate respond at different speeds to excess demand in the product market, the real wage rate and the class distribution of income become endogenous variables of the system through the working of the same multiplier process. As a result, the investment saving gap representing excess of demand in the product market plays simultaneously the role of driving endogenously both the degree of capacity utilization as well as the distribution of income. To represent this distributional adjustment, we introduce in addition to (8) a further equation,

$$dh/dt = \beta\left[I\left(h,z\right) - shz\right], \quad \beta > 0 \text{ or } < 0. \tag{9}$$

Note that the speed of adjustment $\beta > 0$, corresponds to the case of forced saving by the workers, resulting in a lower real wage rate and higher profit share. In turn, it raises saving to help in closing the excess demand gap (Kaldor, 1956). For $\beta < 0$, the real wage increases, as the money wage rate rises faster than the price level in response to excess demand in the product market, and results in a squeeze of the margin, and share of profit. Between forced saving and profit squeeze lies the border-line case, in which the real wage rate remains constant because price and money wage tend to rise at the same proportional rate (Kalecki, 1971). However, this particular case of $\beta = 0$ will not be considered further in the present analysis, insofar as it boils down formally to the case of pure quantity-adjustment in our analysis.[3]

From a formal point of view, the dynamics represented by (8) and (9) are the same, except for their relative speeds of adjustment, represented by the multiplicative terms α and β. Therefore, in equilibrium when the adjustment speeds play no role, the investment saving equality provides only one equation for determining two endogenous variables, z and h.

Faced with this problem of two endogenous variables to be determined by only one equation in equilibrium, many models in the Keynesian tradition have considered in isolation either quantity-adjustment in (8), or distributional adjustment in (9) in order to define the equilibrium configuration from the demand side. However, this amounts to a mode of theorizing in which either the profit share or the degree of capacity utilization has to be treated either as an exogenous variable or, more plausibly, as part of a larger system which provides more equations (and also variables) in addition to the investment-saving equality condition.

However, even relying exclusively on the saving investment dynamics of (8) and (9), we can obtain further insight not merely into the properties of profit- and wage-led growth, but the link that exists between growth and class distribution of income from the demand side (Foley and Michl, 1999).

In this general framework, if $\alpha > 0$, and $\beta > 0$, both z and h would rise endogenously in response to excess demand in the product market. This implies that, in the out-of-equilibrium dynamics, a positive relation holds between z and h. Consequently, the out-of-equilibrium path is profit-led, and entails 'forced saving' by the workers along that out-of-equilibrium path. On the other hand, if $\alpha > 0$, but $\beta < 0$, an excess of investment over saving drives h and z in opposite directions, and the out-of-equilibrium path is wage-led involving 'profit squeeze' on the firms.

This is captured formally by the 'integral curve' obtained by dividing (8) by (9), at non-zero, i.e. out of equilibrium values, without equality obtaining between investment and saving. With the time variable suppressed, this yields the integral curve in the h-z space as,

$$dz / dh = \alpha / \beta, \qquad \alpha > 0, \beta > 0, \text{ or } < 0. \tag{10}$$

Therefore, the direction of the out-of-equilibrium slope (dz/dh) is governed by the sign of β, i.e. whether forced saving ($\beta > 0$) or profit squeeze ($\beta < 0$) takes place in out-of-equilibrium situations.

The stability of the dynamical system (8) and (9) can be examined most generally by considering the function,

$$V = (1/2)\left[I(z,h) - shz\right]^2 \tag{11}$$

Since it is positive definite, and unbounded as $(I - S)$ tends to infinity, stability within a domain is guaranteed by the second method of Liapunov so long as $dV/dt < 0$ (LaSalle and Lefschetz, 1961; Gandolfo, 1996). Differentiating (1) and (3) with respect to time, and substituting from (8) and (9), we obtain,

$$dV/dt = (I - S)^2\left[\alpha(I_z - sh) + \beta(I_h - sz)\right], \tag{12}$$

yielding the condition for stability,[4]

$$\alpha(I_z - sh) + \beta(I_h - sz) < 0. \tag{13}$$

The stability condition (13) can be interpreted economically. Note that eliminating excess demand in the commodity market,

i.e. $(I - S) > 0$ requires, $(dI / dt) < (dS / dt)$; and the opposite case, $(I - S) < 0$ requires $(dI / dt) > (dS / dt)$. Differentiating (1) and (3) with respect to time, and substituting from (8) and (9), inequality (13) can be derived as the product term which satisfies simultaneously both situations covering $(I - S) > 0$, and $(I - S) < 0$ within the relevant domain.

Condition (13) involves two product terms $\alpha(I_z - sh)$ and $\beta(I_h - sz)$. When both these terms are positive, condition (13) is necessarily violated, and the system is unambiguously unstable. These totally unstable cases are ruled out from further discussion for the sake of brevity. On the other hand, when the same two product terms are negative, the system is unambiguously stable. Therefore unambiguous stability requires,

$$\alpha(I_z - sh) \text{ and } \beta(I_h - sz) < 0, \tag{14}$$

i.e.

$$\alpha > 0, (I_z - sh) < 0, \beta > 0 \text{ and } (I_h - sz) < 0 \tag{14.1}$$

or,

$$\alpha > 0, (I_z - sh) < 0, \beta < 0 \text{ and } (I_h - sz) > 0. \tag{14.2}$$

Note that in both these cases of unambiguous stability the Keynesian stability condition (5) is satisfied, but the system can be unambiguously stable with either forced saving or profit squeeze. This dispels the frequently held misconception that the system can be stable only in situations of forced saving due to money illusion or imperfect information.

In (14.1) wage led equilibrium expansion along the *IS* curve occurs according to its slope given by (4), but the system is characterized by profit-led out-of-equilibrium dynamics by the slope of the integral curve (10). Therefore the equilibrium dynamics along the *IS* curve is contradictory to the out-of-equilibrium dynamics of the integral curve (Diagram 1 corresponds to this case).

Contrariwise, in the other case (14.2) of unambiguous stability, equilibrium expansion along *IS* is profit led by (4), but out-of-equilibrium dynamics is wage led by (10).

(Diagram 2 corresponds to this case).

Closer inspection reveals the pattern that in all cases of unambiguous stability (also instability), equilibrium dynamics along the *IS* curve according to (4), and out-of-equilibrium dynamics according to (10) are characterized by contradictory properties of the two regimes. If the equilibrium movement is wage led because h and z are negatively related along the *IS* curve, the out-of-equilibrium dynamics shown by the arrows is profit led as h and z move together along the arrow (see Diagram 1), and vice versa (see Diagram 2).

There are four remaining cases characterized by the ambiguous stability property insofar as the relative magnitudes of the speeds of adjustment enter the stability condition (13) in an essential way. In these cases the stability condition (13) involving the two product terms, $\alpha(I_z - sh)$ and $\beta(I_h - sz)$ are of opposite signs, so that the relative magnitudes of α and β become

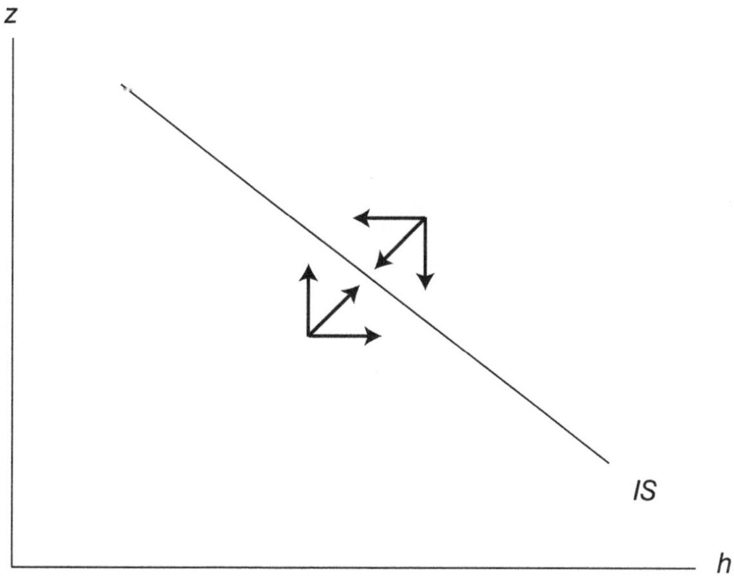

Diagram 1

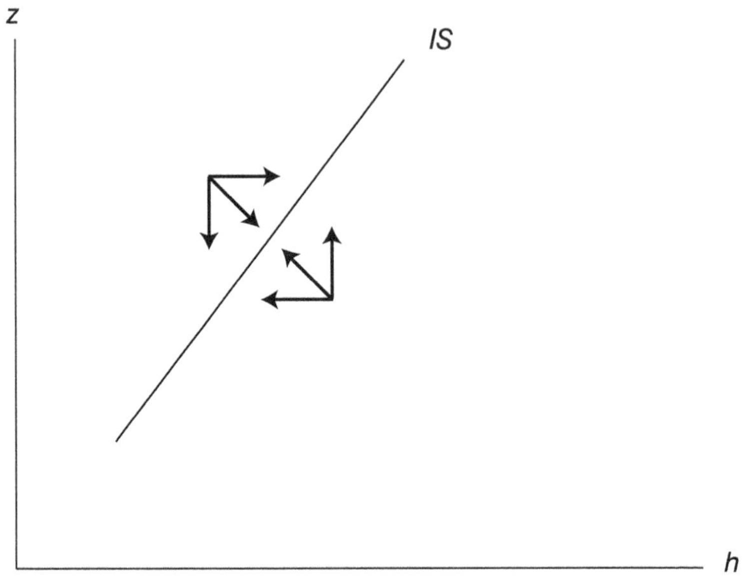

Diagram 2

critical in determining the sign of their algebraic sum. Geometrically, as shown in each case of the following four phase diagrams 3 to 6, the relative magnitudes of the speeds of adjustment control the slopes of the relevant out-of-equilibrium profit – or wage-led trajectories. In the stable cases the trajectories end up on equilibrium rest points on the *IS* curve shown by the solid arrows, while in the unstable cases they move away from the same *IS* curve shown by the broken arrows.

A closer look at the phase diagrams 3–6 of all the ambiguously stable cases reveals a general pattern which can be contrasted against the unambiguously stable (also unstable) cases depicted in diagrams 1 and 2. In diagrams 3–6 the equilibrium expansion along the *IS* curve (equation (4)), and the out-of-equilibrium movement along the arrows (equation (10)) qualitatively belong to the same regime, i.e. both are either profit led (diagrams 3 and 4) or both are wage led (diagrams 5 and 6).

A comparative static exercise would be useful at this point to clarify the relation between the *IS* equilibrium slope condition (4) and the out of equilibrium integral curve slope condition (10). Consider the influence of a boom in the stock market on both investment and saving within the framework of this model. A boom in the stock market represented by a higher value of the parameter λ would usually stimulate both investment and consumption, the latter implying in turn a depressive effect on saving

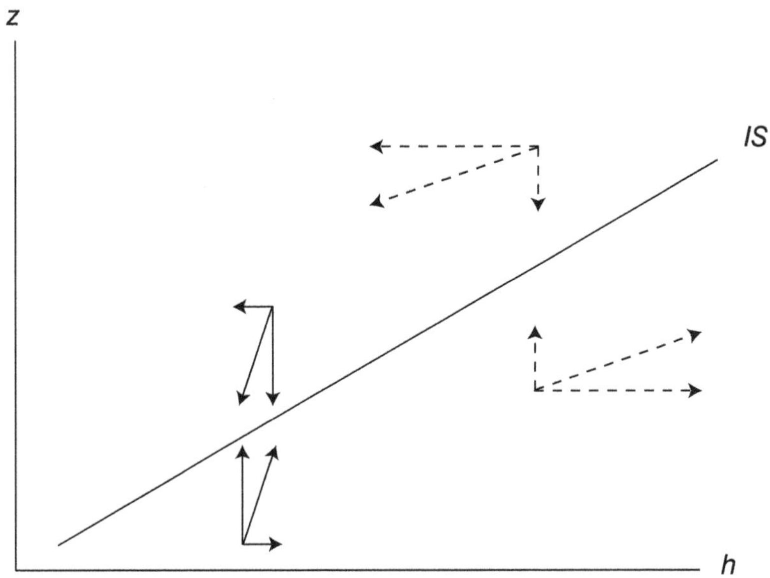

Diagram 3 Profit led by *IS* curve slope (equation (3)) with Keynesian stability satisfied (condition (4)). $\alpha > 0, (I_z - sh) < 0, \beta > 0, (I_h - sz) > 0$.

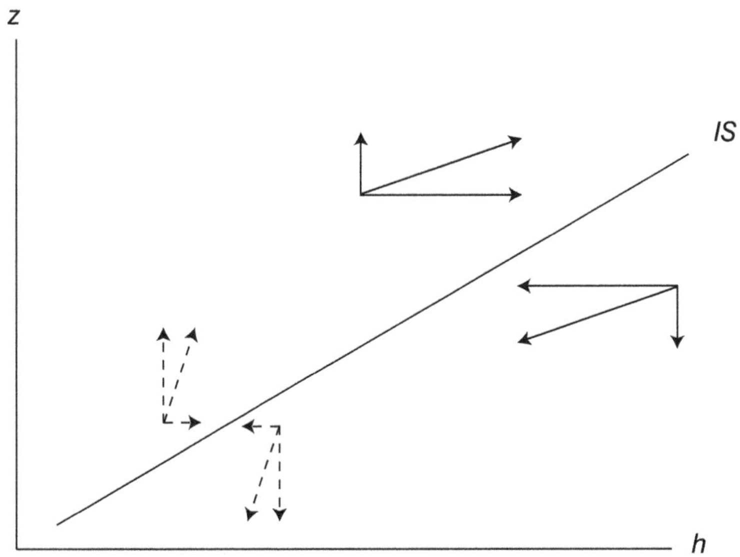

Diagram 4 Profit led by *IS* curve slope (equaton (3)) with Keynesian stability violated (condition (4)). $\alpha > 0, (I_z - sh) > 0, \beta > 0, (I_h - sz) < 0$.

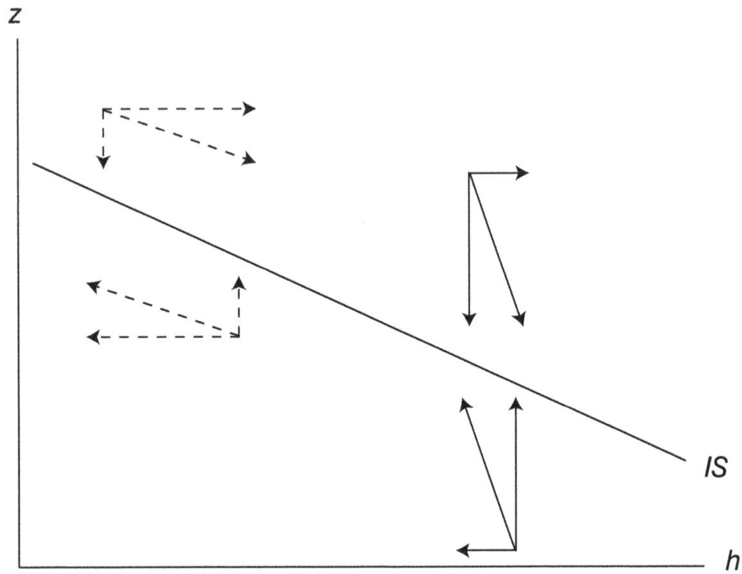

Diagram 5 Wage led by *IS* curve slope (equation(3)) with Keynesian stability satisfied (condition (4)). $\alpha > 0, (I_z - sh) < 0, \beta < 0, (I_h - sz) < 0.$

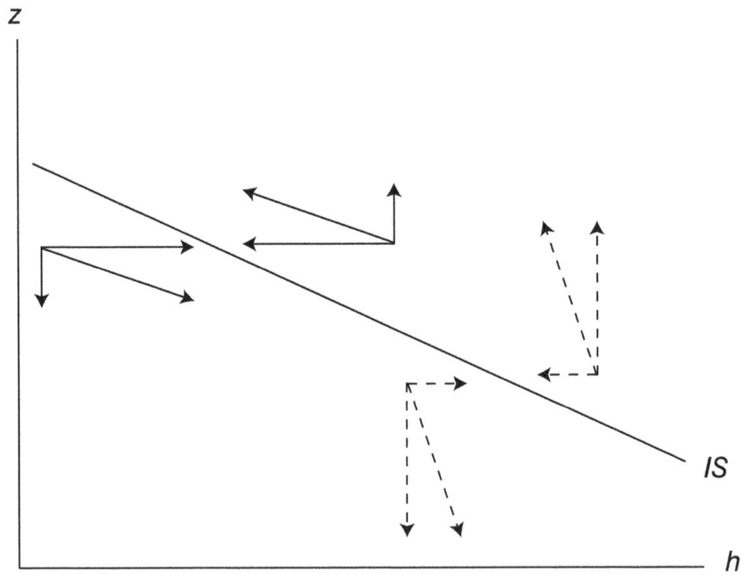

Diagram 6 Wage led by *IS* curve slope (equation (3)) with Keynesian stability violated (condition (4)). $\alpha > 0, (I_z - sh) > 0, \beta < 0, (I_h - sz) > 0.$

propensity (Bhaduri *et al.*, 2006; Maki and Palumbo, 2001). The *IS* equilibrium is written in this case as,

$$I(h, z, \lambda) = s(\lambda)hz, \quad s'(\lambda) < 0 \text{ and } I_\lambda > 0. \tag{15}$$

Totally differentiating (15) and collecting terms,

$$\left[(I_h - sz) + (I_z - sh)(dz/dh)\right]dh = \left[s'(\lambda)hz - I_\lambda\right]d\lambda. \tag{16}$$

In order to consider the stability of the trajectories, we insert in the left-hand side of (16) the out of equilibrium slope of the integral curve from (9) due to a perturbation through λ and simplify to obtain,

$$(dh/d\lambda) = \beta\left[s'(\lambda)hz - I_\lambda\right]/\left[\alpha(I_z - sh) + \beta(I_h - sz)\right]. \tag{17}$$

By assumptions the square-bracketed term in the numerator of the expression on the right-hand side of (17) is negative, and for stable systems, by condition (13) the denominator of (17) is also negative. Consequently the sign of β determines the sign of the expression, and the comparative static result follows, $(dh / d\lambda) > 0$, if $\beta > 0$ (forced saving); but $(dh/d\lambda) < 0$, if $\beta < 0$ (profit squeeze).

In other words, a stock market boom can work for or against the working class. It results in a higher equilibrium profit share under forced saving by the workers, but in a lower profit share in situations of profit squeeze. Although I will not go through other similar comparative static exercises, it should be obvious that we can similarly examine the influence of a higher interest rate or capital inflow in an open economy by specifying their impact on investment and saving.

The generalized Keynesian multiplier mechanism, operating through the adjustment of saving to investment through either capacity utilization or distribution, has also been at the centre of various models of growth inspired by Keynesian ideas. Following Harrod's seminal paper (1939) which focused attention on capacity utilization for examining the instability of adjustment between the warranted and the actual rate of growth, subsequent growth models in the Keynesian tradition tried to provide greater stability to the same adjustment process through the distribution of income (e.g. Kaldor, 1956, 1957; Kahn, 1959; Pasinetti, 1962; Robinson, 1956, 1962; Marglin, 1984).

However, unlike in the present discussion, these demand-led models of economic growth tended to focus exclusively on adjustment through either capacity utilization or income distribution, but not both. A well-known model due to Joan Robinson (1962) is of particular interest in this context.

It focuses on the rate of profit as the central variable in the adjustment of saving to investment. However, because the rate of profit is influenced both by the degree of capacity utilization as well as the distribution of income, it should involve adjustment in both z and h. This is seen from decomposing the profit rate as,

$$r = (R/Y).(Y/Y^*).(Y^*/K) = h.z.q, \tag{18}$$

where Y^* = full capacity output, and q = full capacity output to capital ratio assumed to be a constant in this model due to the Harrod-neutral (1942) nature of technical change and is set at unity for expositional convenience.[5]

Dividing both sides of the investment saving equality by the book value of capital (K), we obtain the 'Cambridge equation' (Hicks, 1965), i.e. $(I/K) = s(R/K)$, where R= total profits, i.e.

$$g = sr. \tag{19}$$

With the equality between investment and saving implied in (19), the warranted rate of growth is equal to the actual rate of growth in Harrod's terminology. The left-hand side of equation (19) represents the warranted rate, interpreted by Robinson as the rate of accumulation desired by investing entrepreneurs. The right-hand side of the same equation represents the actual rate of growth given by the growth of realized saving of the households.

As before, assuming 'business as usual', the ruling rate of profit is viewed by business as a fairly good proxy for the expected rate of profit. This links the rate of accumulation g desired by businessmen to the expected and therefore, actual rate of profit to yield,[6]

$$g = F(r), \quad F'(r) > 0. \tag{20}$$

A higher rate of accumulation, i.e. a higher warranted rate of growth $F(r)$ in relation to the growth in realized savings sr, interpreted as Harrod's actual rate of growth, would create excess demand in the commodity market. This would lead to adjustment in the rate of profit through changes in higher capacity utilization, and/or higher profit share. This is captured by the adjustment equation,

$$(dr/dt) = \xi\left[F(r) - sr\right], \tag{21}$$

Where $\xi > 0$ is the arbitrary constant speed of adjustment.

This one-variable dynamical system is stable provided,

$$F'(r) - s < 0, \tag{22}$$

implying saving is more responsive than investment to changes in the rate of profit r, and is in conformity with the standard Keynesian stability condition (5). Totally differentiating (19) and using stability condition (22), the comparative static result for stable systems follows,

$$(dr/ds) = r/\left[F'(r) - s\right] < 0. \tag{23}$$

This shows that a parametric increase in the propensity to save out of profit (s) weakens the multiplier mechanism in this demand-driven framework to reduce the rate of profit in the new equilibrium.

In order to recast the present analysis with explicit adjustment in h and z, we write separately the adjustment in capacity utilization and profit share as,

$$(dz/dt) = \alpha\left[F(r) - sr\right], \alpha > 0 \tag{24}$$

$$(dh/dt) = \beta\left[F(r) - sr\right], \beta > 0 \text{ or } \beta < 0. \tag{25}$$

Note that both forced saving by workers ($\beta > 0$)) and profit squeeze on the firms ($\beta > 0$) are accommodated in (25), as explained earlier in connection with equation (9).

Proceeding in the same way, and forming the Liapunov function corresponding to (11),

$$V = (1/2)\left[F(r) - sr\right]^2, \tag{26}$$

and using (23) to (26), the condition for stability is obtained as,

$$(dV/dt) = q\left[F(r) - sr\right]^2 \left[F'(r) - s\right](\alpha h + \beta z) < 0, \tag{27}$$

i.e.

$$\left[F'(r) - s\right](\alpha h + \beta z) < 0. \tag{28}$$

Under the assumption of forced saving by the workers ($\beta > 0$), it follows from (28) that the earlier stability condition (21) is sufficient.

However, an interesting additional possibility emerges in this more generalized framework by allowing for profit squeeze ($\beta < 0$). It is possible that the standard Keynesian stability condition (22) is violated, and still the system remains stable through satisfying (28), provided $[F'(r) - s] > 0$, and $(\alpha h + \beta z) < 0$. This would imply that the comparative static result (23) is reversed in this case. In order to see why this happens, we consider first the out of equilibrium dynamics.

From the 'integral curve' obtained by dividing (25) by (24) at non-zero values of (I-S), i.e. for out of equilibrium values, we obtain the out of equilibrium slope of adjustment in the *hz* plane as,

$$(dh/dz) = (\beta/\alpha). \tag{29}$$

From total differentiation of (18) and (19), we obtain,

$$[F'(r) - s][h.dz + z.dh] = r.ds. \tag{30}$$

Inserting (29) in (30) and simplifying, we arrive at the formalism necessary for two comparative static results,

$$(dh/ds) = \beta r / [F'(r) - s](\alpha h + \beta z), \quad \beta > 0 \text{ or } \beta < 0. \tag{31.1}$$

$$(dz/ds) = \alpha r / [F'(r) - s](\alpha h + \beta z), \quad \alpha > 0. \tag{31.2}$$

Provided stability condition (28) is satisfied, it follows from (31.2) that in the new equilibrium at a higher saving propensity, capacity utilization z is lower due to the weakening of effective demand through the multiplier, and from (31.1) profit share would also be lower under forced saving ($\beta > 0$). However, under profit squeeze ($\beta < 0$), profit share would fall at a slower rate as aggregate demand slackens due to the higher saving propensity. If this relatively favourable impact on profit share is sufficient to outweigh the negative impact of lower capacity utilization, the equilibrium rate of profit would rise despite an increase in the saving propensity, reversing the comparative static prediction in (23).

Economic models of such simplicity are not meant to be taken literally. They are more like cartoons made to draw our attention to particular features of a far larger and complex reality. Aggregative Keynesian demand determined output models focus sharply only on disequilibrium in the commodity market, leaving out of the picture some crucial aspects like inventory adjustment, the labour and the money market (cf. Chiarella and Flaschel, 2000). And yet models of such simplicity might highlight some particular problems inherent in the process of capitalistic growth. For instance, at the equilibrium rate of growth of this model the sales expectations of business are satisfied, as the commodity market is cleared through the equality between investment and saving. As a result, it also generates an equilibrium rate of profit, and an equilibrium growth path that satisfies the sales expectations of business. And yet, it is a growth path that need not guarantee full employment; indeed the unemployment rate might even continue to rise, if the natural growth rate of the labour force exceeds the growth rate considered as equilibrium by business. In other words, what appears as an equilibrium state of affairs from the point of view of business might represent at the same time a disequilibrium state of affairs for the workers facing growing unemployment. The message of such a model is clear. The notion of a commodity market clearing equilibrium in this economy is multi-faceted; it might appear different depending on the point of view of the class which considers it. One might try to widen the notion of equilibrium by incorporating more markets. However, in some markets the two contending parties might have a different notion of what the equilibrium state of affairs should be, even if one party or the other does not have the power to alter it.[7] Macroeconomic equilibrium in this sense is not necessarily a neutral notion, and one usually needs to look deeper into the balance of power of the classes to make a judgment about its sustainability.

3 Lecture III: A model of endogenous growth driven by intra- and inter-class competition

In a growing economy the problem of effective demand is best introduced in a formal way, by allowing for the possibility of disequilibrium between the growth rates of investment and of saving. To capture this formally, suppose initially the economy is in a state of equilibrium growth. This implies that the commodity market is cleared, and the equilibrium ratio of investment (*I*) to saving (*S*) remains at unity, $(I^* / S^*) = 1$ This equilibrium ratio defines the locus of all possible commodity market clearing positions, and the equilibrium is maintained so long as the rate of growth is the same for investment and for saving. To develop the analysis farther along these lines, we postulate a power function,

$$Y = A[I / S]^{\alpha}, \quad \alpha > 0, A = Y^* > 0. \tag{32}$$

Note that for all equilibrium values of the ratio $\left(\frac{I}{S}\right) = 1, \left(\frac{Y}{A}\right) = 1$. Thus, *A* represents the locus of all the commodity market clearing equilibrium values. Writing $g_A = g_{y^*}$, for commodity market clearing growth rate of *Y*, logarithmic differentiation of (32) yields $(g_y - g_{y^*}) = \alpha(g_I - g_s)$.

As the formulation above shows, the deviation in the growth rate of output g_y from its market clearing growth rate g_{y^*} is induced by the disequilibrium between the growth rates in *I* and *S*, which in turn triggers off an adjustment in *Y*. Assuming savings as an increasing function of income, this would raise g_s sufficiently through higher g_Y until the right-hand side reaches zero. Approximating over continuous time, we write

$$\frac{dg_Y}{dt} = \alpha(g_I - g_S) \tag{33}$$

where α is some arbitrary positive speed of adjustment.

An investment (demand) function different from the saving function has to be introduced to capture the possibility of a divergence between their respective rates of growth. For neither the investment nor the saving function, a commonly accepted formulation exists. However, for simplicity of exposition it would suffice to consider a regime of static expectations, in which investment depends positively both on the current level of output as a predictor of the future state of demand as well as on labour productivity as a predictor of expected profitability. Note expected profit margin increases through higher labour productivity, only if productivity increases without a compensating increase in the real wage rate. Consequently, productivity (X) could be treated as an indicator of profitability, if business takes the real wage as a given datum beyond its control. On these assumptions, the investment function is specified as, $I = I(\overset{+}{Y}, \overset{+}{X})$, with the sign of the partial derivative with respect to the relevant variable shown above it. Simple manipulation converts it into the growth rate of investment as

$$g_I = \eta_y g_y + \eta_x g_x \tag{34}$$

where, g_j = the growth rate of variable j, Y = output level, X = labour productivity level, and, η_y and η_x are positive partial elasticities of investment with respect to output and investment respectively. In its simplest specification, saving is treated as an increasing function only of income,[8] so that

$$g_s = \varepsilon_y g_y, \tag{35}$$

Where ε_y is the positive elasticity of saving with respect to income.

Inserting (34) and (35) in (33), we obtain

$$dg_y / dt = \alpha\left[(\eta_y - \varepsilon_y)g_y + \eta_x g_x\right]; \quad \eta_y, \varepsilon_y, \eta_x > 0. \tag{36}$$

The supply of labour force is assumed to grow at a steady exponential rate, n. This ignores many economic and sociological factors that tend to affect the supply of labour, e.g. economic migration, flexible participation of married women in the labour force, absenteeism influenced by the state of public health, as well as cultural conventions that encourage or restrict participation. With all such complications about the labour supply set aside, any given rate of unemployment would remain constant so long as employment grows at the same rate as labour supply, i.e. $g_L = n$.

The constant rate of unemployment might be considered as an 'equilibrium position' of the labour market with very different economic interpretations. It ranges from the Marxian 'reserve army' of labour (Marx, 1867) to the 'natural' rate of unemployment (Friedmam, 1968), and the non-accelerating inflation rate of unemployment (NAIRU). The particular interpretation favoured depends on the out-of-equilibrium dynamics postulated. The simplest postulate in the present context is to suggest that a discipline on labour is imposed by the capitalists though maintaining a constant fraction of the labour force unemployed. This would be the reserve army of labour, broadly in conformity with the Marxian view, according to which this army is maintained through the continuous introduction of various labour-saving devices. This would raise labour productivity in accordance with the degree of tightness of the labour market, and keep a constant fraction of the labour force continuously unemployed. This view of labour productivity growth, induced by inter-class conflict in the labour market is represented formally as

$$(dg_x / dt) = \beta(g_L - n) = \beta(g_y - g_x - n), \quad \beta > 0 \tag{37}$$

where, by definition

$$g_y = g_x + g_L. \tag{38}$$

Equations (36) and (37) depict a dynamical system in which the growth rate of output g_Y is governed by the growth of demand in the product market according to Keynes, while that of labour productivity g_X is influenced by inter-class conflict according to Marx. Despite its simplistic nature, this provisional model of growth along the Marx–Keynes tradition highlights how labour productivity growth endogenously driven by inter-class conflict can interact with output growth through effective demand. To analyse the formal properties of this model, note from (36) and (37),

$$dg_y / dt = 0 \text{ implies } g_y = \theta g_x \text{ where } \theta = \left[\eta_x / (\varepsilon_y - \eta_y)\right]; \tag{39}$$

and

$$dg_x / dt = 0 \text{ implies } g_y = n + g_x, \quad n > 0. \tag{40}$$

It follows from (39) and (40) that no positive configuration of equilibrium growth rates in output (g_y^*), labour productivity (g_x^*), and employment (g_L^*) is possible unless

$$\theta > 1, \text{ i.e. } \eta_x > (\varepsilon_y - \eta_y) > 0. \tag{41}$$

The last inequality in (41) is a mere restatement of earlier stability condition (5) for output adjustment in the Keynesian system. The equilibrium rates of growth computed from (39) and (40) are

$$g_y^* = n\theta/(\theta - 1),\ g_x^* = n/(\theta - 1) \text{ and } g_L^* = n \tag{42}$$

with all the growth rates in (42) positive provided that $\theta > 1$, i.e. inequality (41) is satisfied. This is exhibited in Diagram 7, where the line AB corresponding to $(dg_x/dt) = 0$ in (40) passes through $(-n, 0)$ and $(0, n)$ with a positive slope of unity. For a positive solution in g_x and g_y to exist at E, the ray through the origin OC corresponding to $(dg_y/dt) = 0$ in (39) must have a slope greater than unity implying that $\theta > 1$. However, Diagram 7 on the phase plane also suggests that the solution is a saddle point, which cannot generally be considered stable. This could be confirmed by checking whether the relevant Jacobian matrix of the dynamical system (36) and (37) evaluated at equilibrium has a negative trace, and a positive determinant to ensure stability. The trace, $-\alpha(\varepsilon_y - \eta_y) - \beta$ is negative under (41), but the condition for a positive determinant, i.e. $\alpha.\beta(\varepsilon_y - \eta_y - \eta_X) > 0$, is seen to violate the same condition (41). This implies that a positive solution for g_y^* and g_x^* in (42) either does not exist or is an unstable saddle-path.[9]

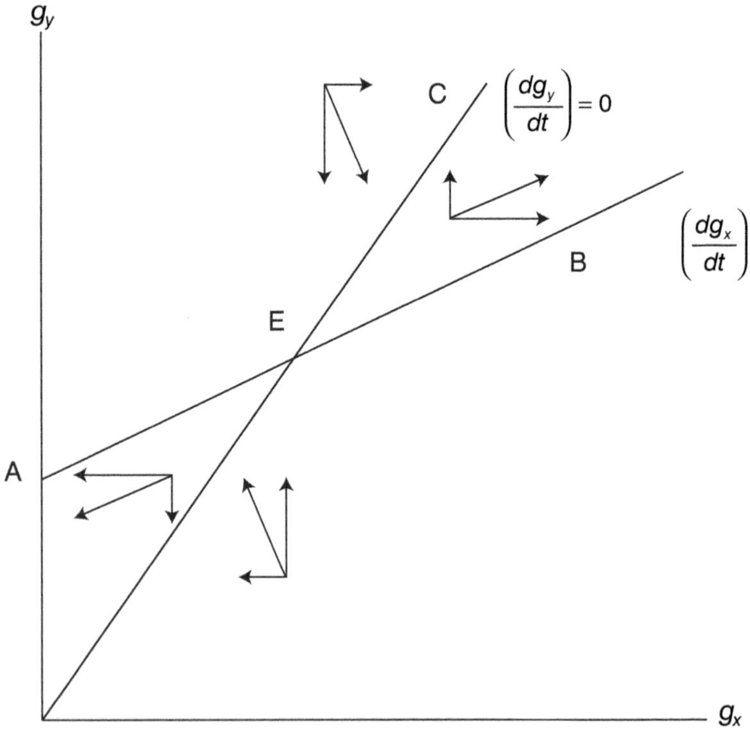

Diagram 7 Notes: OA = n; slope of AB = tan 45°; slope of OC = tan $\theta(\theta > 1)$; coordinates of equilibrium at E $\left(g_x^*; g_y^*\right)$ can be derived from equation (42)

Through various modifications especially of the investment function, this miniature model incorporating some features of Keynesian effective demand and Marxian reserve army of labour maintained through endogenous growth in labour productivity can be rendered less unstable. However, we would not pursue this track farther, because the acceptability of any particular investment function would in any case remain controversial. It might be more fruitful instead to focus attention on the lacunae arising from the over-simplistic specification of the endogenous productivity growth equation (37).

The specification of productivity growth in (37) ignores altogether intra-class competition over market shares among rival capitalist firms in driving the diffusion of productivity growth in a competitive or monopolistic market structure. And yet, this view of technological competition among rival firms is shared by almost all schools of thought. In the classical view of competition, the price level is driven down continuously as innovations get diffused across the economy to lower production cost. Schumpeter in his early writings had argued too that entrepreneurial profit from each particular round of innovation is transient. It would tend to be competed away through the diffusion of technology among numerous small firms with free entry or exit. Thus, the economy would tend towards an equilibrium in which prices tend towards their lower production costs lowered continuously through innovations. Only 'normal' profit, operating through the 'law of cost', as Schumpeter called it, would exist in equilibrium until the system is disturbed again by a fresh round of innovations (Schumpeter, 1961: Ch. 5).

The diffusion of technological knowledge from the innovating firm to its rivals implies that productive knowledge is a non-exclusive public good. It generates positive externalities and increasing returns. However, the speed of diffusion of technology would depend, among other things, on the fierceness of competition, the cost of adopting the new technology, and especially the degree of restrictiveness imposed by the particular legal regime of property rights.

The process of technology diffusion might be captured formally by imagining that there are numerous small firms in the economy with free entry and exit. Inverting the image of the Ricardian theory of differential land rent, we postulate that the industrial economy is subject to increasing returns due to the public good character of productive knowledge. This means that labour is more productive in the best-practice firm at the frontier or 'margin' of technological knowledge compared with the average productivity of labour in the rest of the economy. Since the marginal firm at the frontier of knowledge is one among numerous small firms, it has a sufficiently low weight in the total output produced in the economy. Consequently its higher productivity raises only negligibly the average level of productivity in the economy.[10]

Assuming intra-class competition among the firms along the lines suggested by the classical economists, the marginal firm at the technological frontier lowers its price in accordance with its lower production cost to capture a higher share of the market. Thus, the cost differentials among the firms are assumed not to affect their respective profit margin per unit of sale, but get converted into a competition among them for increasing the relative volume of sales through higher market share. This price-setting behaviour puts a general downward pressure on the price level, because the technologically advanced firm by lowering its price emerges as the price leader under classical competition.[11] The numerous remaining firms follow the price-leader in order to survive in a competitive market by lowering their costs and prices through adopting gradually the new technology over time. Thus technological competition results exclusively in price competition with all firms maintaining the same mark-up on their costs, while fighting over market shares. The benefit of cost-reducing technology is passed on to the consumers in this process without any higher profit margin for even the innovating firm. This process of reduction in costs and prices with increasing returns through the diffusion of technology might be formalized as a discrete process. Let P_t^f = price set by the firm at the frontier of technology at time t, P_t = the average ruling price (of the numerous other firms) in the economy. Assuming a uniform mark-up (k), and money wage (w), the mark-up prices with labour as the only variable cost are

$$P_t^f = k.w.(dL/dY) \text{ and } p_t = k.w.(L/Y) \tag{43}$$

where,

$P_t^f < P_t$ because $(dL / dY) < (L / Y)$ due to increasing returns, w = money wage rate, and L = employment, Y= output, and $(Y/L) = x$, labour productivity. If technology diffuses at a uniform rate, and it takes τ periods for each innovation to be diffused completely, then the adjustment in the price level in each period is $(P_t^f - P_t)/\tau$. Therefore, the average price in the next period becomes

$$\begin{aligned} &P_{t+1} = P_t + \left[\left(P_t^f - P_t\right)/\tau\right], \text{ or} \\ &(P_{t+1} - P_t) = \Delta P_t = (dP/dt) = \lambda(P^f - P), \quad \lambda = (1/\tau). \end{aligned} \tag{44}$$

The speed of adjustment λ indicates the ease with which innovation gets diffused in the economy. For example, given the market structure and other parameters of competition, a tighter regime of intellectual property rights would make the value of λ smaller, reducing in turn the speed of price reduction in the economy.

Using (38), (43) and (44) we obtain on simplification the decline in the rate of average price level in percentage terms as

$$g_p = -\lambda(g_x / g_y), \quad g_y \neq 0. \tag{45}$$

Note that the elasticity of employment with respect to output, i.e. $[(dL / L)/(dY / Y)]$ is rewritten in (45) in terms of the growth rates by the chain rule of differentiation to capture the dynamic nature of increasing returns (Kaldor, 1989).

Equation (45) captures the consequence of technological competition among rival firms in terms of a secular downward pressure on the price level, generated by increasing labour productivity resulting from dynamic increasing returns. It is measured here by the elasticity of employment with respect to output over time, i.e. (g_L/g_Y). The smaller is this ratio of employment elasticity of output, the larger would be the elasticity of labour productivity with respect to output from definition (38), and the larger is the consequent reduction in unit cost and the price level. The benefits of price reduction through technical progress get passed on ultimately to the consumers as firms contest over market shares in this classically competitive market structure.

However, while price reduction takes place according to equation (45), the behaviour of the real wage rate depends also on how the money wage rate changes in the labour market through inter-class competition. The inter-class competition in the labour market can be brought into the picture through two alternative routes. The first route is to assume a standard 'Phillips curve', postulating a stable negative relation between the growth in money wage and the unemployment rate. If $g_L = n$, the percentage of unemployment remains constant, and the corresponding growth rate in money wage rate is determined through the traditional Phillips curve. Accordingly, the growth rate of money wage, μ is determined at some specified constant rate of unemployment, u^*, i.e.

$$g_\vartheta = \mu \text{ at } u = u^*, \text{ implying } g_L = n. \tag{46}$$

Therefore the increase in real wage υ is given from (45) and (46) as,

$$g_\vartheta = \mu + \lambda(g_X / g_Y). \tag{47}$$

If labour productivity (g_X) continues to rise at a rate different from that of the real wage rate over time, the share of wages in income would also continue to rise or fall, becoming ultimately incompatible with steady state growth. Therefore labour productivity growth might be postulated to adjust in such a manner as to keep the wage share constant in the long run giving technical progress the appearance of 'neutrality'. This yields an alternative adjustment mechanism for productivity growth in place of (37) as,

$$dg_x / dt = \beta[g_v - g_x] = \beta[\mu + \lambda(g_X / g_Y) - g_X]. \tag{48}$$

Note that the tendency towards a constant wage share is not postulated here as a technological datum. Wage share tends to be constant as an outcome of inter-class competition, and this tends to make technical progress appear as Harrod-neutral (Harrod, 1942; Uzawa, 1961). This contrasts sharply against the usual practice of neo-classical models of endogenous growth. Unlike in those models the constancy of the share of wages over the long run is not obtained through any ad hoc assumptions like Cobb–Douglas technology, supplemented by the human capital approach. Instead, the co-evolution of rising labour productivity through technological progress, and rising real wage rate are viewed in (48) as two sides of the same economic process, driven endogenously by intra- as well as inter-class competition, leading to a tendency towards a rough constancy of the wage share over time. In some ways, this view of technological progress comes close to the Smith–Marx vision. However, in one important respect it differs from the classical, especially the Marxian view. In the present formulation, technological development tends to keep the wage share, but not the real wage rate, constant over the long run.[12]. It also emphasizes the two-sided role of technology in raising the real wage rate through cost competition among rival firms on the one hand, and not allowing the wage share to rise on the other. While the wage share is an outcome of the race between new innovations and rising real wages within the framework of this model, more complex adjustments could be visualized to permit short-term fluctuations in wage share with a tendency towards constancy in the longer run.

Equations (36) and (48) define a dynamical system in which Keynesian effective demand interacts with technological change propelled endogenously by intra- and inter-class competition. However, the model imposes rather arbitrarily through equation (46) a constant rate of unemployment.

From (36), (46) and (48), the equilibrium configuration is computed as

$$g_L^* = n;\ g_X^* = n/(\theta - 1) \text{ and } g_Y^* = n\theta/(\theta - 1), \tag{49}$$

with the further restriction from (41), $\theta = \eta_X / (\varepsilon_Y - \eta_Y) > 1$.

In contrast to the main result of the standard neo-classical model (Solow, 1956), the relationship between the equilibrium growth rate of output $g*_Y$ and the exogenous rate of the labour force n is seen to be quite flexible in (49). It is influenced by the investment and saving behaviour of the economy;

subject to inequality $\theta > 1$, g_Y^* would always lie above n. Moreover, for any given value of n, the smaller is θ the larger is the equilibrium growth rate g_Y^*. For instance if $\theta = 1.1$, the growth rate of output is 11 times higher than the exogenous growth rate of population n; on the other hand if $\theta = 5$, it is only 1.25 times higher. Since θ is determined by the investment and saving behaviour of firms and households respectively, the long-run steady state growth rate is viewed in this framework as the consequence of human agency. As a result it could vary widely across countries, even with the same exogenous growth rate of population.

Carrying out routine computations and evaluating the trace (T) and the determinant (D) of this dynamical system around the equilibrium characterized by (49), it can be checked that

$$D = \alpha\beta(\varepsilon_Y - \mu_Y) > 0 \tag{50}$$

in view of (41), and

$$T = -(\varepsilon_Y - \eta_Y)[\alpha + \beta / n.\eta_X] - \beta[1 - (\lambda / n)]. \tag{51}$$

A sufficient condition for the negativity of T is given by,

$$n > \lambda \tag{52}$$

This last inequality has an interesting economic implication. Intra-class competition among firms is beneficial, driving the diffusion of technology, but only up to a point. Too much competition that raises the value of λ sufficiently to violate inequality (52) would destabilize the system. From this point of view intra-class competition among the firms operates like a double-edged weapon. On the one hand, it benefits all consumers through reduction in the costs of production and prices brought about by the diffusion of technology, raising in turn the real wage rate. On the other hand, however, the same process of rising real wage rate puts increasing pressure on the firms to increase labour productivity at a still faster rate implanting a source of instability into the system.

Although a stable Phillips curve relationship has been used in the preceding formulation, its general validity has often been called into question, particularly as it failed to provide an explanation of the phenomenon of 'stagflation', the co-existence of high inflation and unemployment in the mid-1970s. It has become more common since then with some supporting evidence to reformulate the argument in terms of a 'wage curve', relating the percentage change in the money wage rate to the change in the unemployment rate (du / dt) (e.g. Blanchflower and Oswald, 1994, 1995; Card, 1995). The elasticity of the wage curve, relating the percentage change in wage to that in unemployment yields $-b = (dw/dt/w)/(du/dt/u)$, where b is the absolute value of the elasticity.

It can be rewritten as

$$-b = (dw/dt/w)/\left[u/(n-g_t)(1-u)\right]. \tag{53}$$

Expression (53) is useful in showing explicitly that the magnitude of the elasticity would normally be different at different rates of unemployment. However, for expositional convenience, we would assume a constant elasticity form relating the wage level to the employment rate, i.e.

$$w = j(L/N)^m, \quad m > 0, \tag{54}$$

where j could be interpreted as representing the full employment wage rate. This hypothetical full employment wage would normally be expected to rise over time with rising labour productivity.

From the logarithmic differentiation of (54), the growth rate of money wage becomes

$$g_w = g_j + m(g_L - n). \tag{55}$$

Consequently the rate of growth of the real wage rate g_v is given from equations (45) and (55) as

$$g_v = g_w - g_p = g_a + m(g_l - n) + \lambda(g_x / g_y). \tag{56}$$

If labour productivity g_X continues to rise at a rate different from that of the real wage rate, the share of wages in income would also continue to rise or fall over time, ultimately becoming incompatible with a steady state. Therefore, driven by inter-class conflict over distribution, labour productivity growth might be assumed to adjust to keep pace with rising real wage rate. The consequent tendency towards a constant wage share through adjustment in labour productivity growth is captured by

$$\begin{aligned}(dg_x/dt) &= (g_v - g_x) = \\ &\left[g_j + m(g_Y - g_X - n) + \lambda(g_X/g_Y) - g_X\right].\end{aligned} \tag{57}$$

which now replaces the previous equation (48).

The equilibrium of dynamical system (36) and (57) is given by

$$g_X^* = [\theta(g_j - mn) + \lambda] / \theta[1 - m(\theta - 1)],$$
$$g_Y^* = [\theta(g_j - mn) + \lambda] / [1 - m(\theta - 1)]$$
$$\text{and } g_Y^* - g_X^* = g_L^* = \qquad (58)$$
$$[\theta(\theta - 1)\{(g_j - mn) + \lambda\}] / \theta[1 - m(\theta - 1)].$$

The stability of the steady state equilibrium described by (58) requires the relevant Jacobian matrix evaluated at equilibrium to have a negative trace (*T*), and positive determinant (*D*), i.e.

$$T < 0, \text{ implying}$$
$$\alpha(\varepsilon_y - \eta_y) + \beta[(1 + m) - (\lambda / g_y^*)] > 0 \qquad (59)$$

$$D < 0, \text{ implying } \alpha\beta(\varepsilon_Y - \eta_Y)[1 - m(\theta - 1)] > 0. \qquad (60)$$

In view of (41), the determinant condition (60) would be satisfied if

$$\theta < (1 + m) / m. \qquad (61)$$

Note from (54) and (55) that the weaker is the inter-class conflict between labour and capital resulting in a weaker wage response to employment growth, measured by the elasticity *m*, the less restrictive becomes condition (61). As a result, the system is more likely to be stable for weak inter-class conflict with relatively low values of *m*, provided the trace condition (59) is also satisfied. Substituting the value of g_Y^* from (58), the trace condition is seen to be satisfied sufficiently in view of (41) and (61) if

$$(g_j - mn) > \lambda, \qquad (62)$$

whereas from (58) and (60),

$$g_Y^* > 0, \text{ requires } (g_j - mn) > (\lambda / \theta). \qquad (63)$$

Given condition (41), inequality (62) is stricter than (63), and becomes the operative inequality.

An interesting economic implication of (61) and (62) follows. Given the state of intra-class competition among rival capitalists captured by the value of the technology diffusion parameter λ, inequalities (61) and (62) imply that a relatively low value of m indicating weaker inter-class competition between capital and labour, would tend to make the system stable. And, given the value of m, stronger intra-class competition with a higher value of λ would tend to be destabilizing for the system. However, while less acute inter- as well as intra-class competition tend to be stabilizing for the steady state growth path, it can also be seen from (58) that the growth rates of both output g_y^* and productivity g_x^* tend to be stimulated by more intense intra-class competition among rival capitalists. Again, as observed earlier, the competition among rival capitalist firms continues to operate like a double-edged weapon; while stimulating growth, it tends to destabilize it (see Bhaduri, 2006 for elaboration).

Although the economic system depicted above seems capable of a stable configuration of steady, positive growth rates in output, productivity and employment given by (58), it is in fact only a quasi-steady state because the unemployment rate (u) keeps changing, and, with the unemployment rate increasing or decreasing continuously, the quasi-stationary state might not be sustainable for long. Thus, contrary to the comfortable neo-classical story of full employment growth, this model shows that even in the long run, there is no automatic tendency in the market economy to keep the labour market either at full employment or at some constant, so-called 'natural' rate of unemployment. And, this happens despite the fact that the commodity market clears along the steady growth path implying all sales expectations of business are satisfied, and the share of wages also tends to remain constant.

The general message emphasized in the earlier model appears even more starkly in the present one. With effective demand brought into the analysis of growth, the notion of economic equilibrium becomes considerably more restrictive. As seen in these models, continuous commodity market clearing equilibrium, might be incapable of clearing simultaneously the labour market, even if in this more elaborate model the distribution of income gets stabilized through what appears as 'neutral' technical progress. The neutrality of technical progress emerges endogenously as an outcome of intra- and inter-class competition. While technical progress and productivity growth chase one another in a race that is a characteristic feature of capitalist dynamics, the possibility of unemployment as its other intrinsic feature is also present in this model. The demand-driven growth rate of output with endogenous labour productivity growth results in the growth rate of employment as a residual, which may be incapable of ensuring full employment; it may even result in jobless growth or even a growing rate of unemployment (see condition 41). In other words, economic equilibrium, like beauty, is often in the eyes of the beholder! What appears as a tranquil equilibrium for the 'captains of industry' with their sales and profit expectations satisfied, might at the same time appear to the workers as a disturbing situation of disequilibrium, characterized by growing unemployment. It is a moot question how long any such equilibrium with rising unemployment rate could be sustainable. One could ask a similar question also of the 'natural' or NAIRU rate of unemployment. Even if that rate is constant, would it be sustainable, if it is very high?

The message concerning jobless growth could be softened with suitable modifications of some of the assumptions made in this model. For instance, merely as a matter of expositional convenience, the elasticity (m) relating the rate of growth of money wage rate to the growth in the rate of employment was assumed constant in equation (53). However, this elasticity might plausibly be expected to be lower at a lower rate of employment due to the weaker bargaining power of the workers. By reducing the growth rate of real wage at higher unemployment rate, this might reduce the pressure on the growth rate of labour productivity, but without a corresponding decline in output growth. This would exert in turn a stabilizing influence on the labour market through raising the growth rate of employment.

Finally, a difference in perspective between the model of the last lecture and this one should be noted. Unlike the last lecture in which we considered the relative movement of the price level and the money wage rate driven by excess demand in the commodity market as the main determinant of the class distribution of income, the present model views it as the outcome of a contest over the longer run between the growth rate of the real wage, and that of labour productivity. Consequently, endogenous neutral technical progress emerges as an outcome of this contest. From this point of view the neutrality of technical progress is a consequence not of technology, but of intra-class competition among rival firms which raises the real wage rate to put pressure on profit share, and inter-class conflict over distribution resolved partly through rising labour productivity. As a result the share of wages in income tends to be relatively constant over the longer run. The steady state properties in any case would remain largely unaltered, as distribution of income attains constancy in the steady state through the appearance of neutrality in technical progress. However, focusing exclusively on the steady state without due consideration given to out of steady state behaviour can be misleading. In particular, as discussed in the last lecture, the impact of changes in the distribution in out of equilibrium situations should be incorporated as an extension of this model. However, for expositional simplicity, we focused only on a case in which the influence of income distribution on saving is left out. As mentioned earlier, this or other simplifications inevitable in any stylized economic model can be misleading unless it is always remembered that they are like cartoons. They are deliberately simple with exaggerated features, meant to draw attention to particular features of capitalistic growth. Yet, they can be telling like political cartoons if they are related meaningfully to particular aspects of reality. The final lecture will explore how our model might connect with at least some aspects of reality.

4 Lecture IV: Model and reality – a summing up

Let me begin by asking the most awkward question one could ask of an aggregative model of the kind outlined above: how seriously should we take it? It hardly needs saying that it need not be taken too literally. The usefulness of a model of this kind lies partly in its ability to bring to attention certain aspects of the process of economic growth that are important in our view but remain underemphasized in the discussion on the subject. A stylized model is a suitable vehicle for explaining how such omissions may be misleading. At the same time, the usefulness of the model is also to be judged by its own omissions, i.e. what is left out of the model, deliberately or otherwise.

The central feature of the model presented here has been to indicate how endogenous forces drive the growth rate of both output and labour productivity, the former through expanding aggregate demand or the size of the market, and the latter through various forces of inter and intra-class competition. Our emphasis on the role played by aggregate demand or the size of the market in the process of economic growth has been necessary, mainly because mainstream neo-classical growth models increasingly neglect this aspect; and therefore cannot deal with the problem of unemployment in a growing economy. In this framework, the exogeneity or endogeneity of technical progress is treated also as a purely supply side phenomenon, as if generation of technical knowledge is synonymous with its adaptation and diffusion (Solow, 1957). One special feature of our formulation was to point out how that way of dealing with technical progress makes it devoid of its social content without attention being given to the intra-class competition and inter-class conflict in which especially the diffusion of new technology is embedded. Therefore, with their exclusive pre-occupation with the supply side, these models fail to realize how the analysis of capitalistic growth would become different in two significant respects, once the problem of deficiency in aggregate demand, as well as the existence of forces of intra- and inter-class competition that shape endogenously the nature of technical progress are recognized. As the model presented here would make clear, there is no automatic mechanism in a market economy to guarantee continuous full employment during the process of growth in the presence of the problem of deficient aggregate demand. Second, along with demand-determined output, when productivity growth is influenced significantly by the endogenous forces of inter- and intra-class competition, employment growth emerges as a residual variable. In that case full employment growth is far from certain. We point out how not only unemployment but jobless growth, even a rising rate of unemployment become distinct possibilities in a growing economy. It may be objected that this type of scenario is not compatible with the steady state. However, this is just another way of saying that steady growth with full employment or even a constant rate of unemployment may not necessarily exist in a free market economy. This main message of our growth model with demand-determined output, and competition-driven productivity growth needs to be contrasted against the happy land of neo-classical growth theory where full employment is simply assumed as a long-run property of capitalism. And, by assuming continuous full employment, neo-classical supply side models imagine a world that could be dangerously misleading for the circumstances in which we often find ourselves in the real world.

Ignoring the problem of aggregate demand in the context of long-run economic growth is misleading also in another important respect. The conventional neo-classical wisdom is that growth theory is concerned with the potential, rather than the actual output path (cf. Solow, 2000: 184). This position is logically indefensible. To assert that short-term fluctuations take place around some given trend of potential output which in turn is not affected by these fluctuations is unsound. It was mentioned in the very first lecture that actual employment and output behaviour would often tend to leave a persistent effect of hysteresis on the potential growth path of output. Take, for example, the idea that the trend rate of potential output is derivable from actual output adjusted by the rate of unemployment. This would be acceptable if potential labour productivity remained unaltered irrespective of the unemployment rate. And yet, the level of potential labour productivity is known to be influenced strongly by the actual growth path of output and employment through a variety of factors, like on-the-job skill that the unemployed labour force could not be acquire, the cyclical fluctuations in productivity that affects expectations, or the pace of endogenous productivity growth linked to output growth emphasized in particular by Verdoon and Kaldor. Opinions may differ about the importance of these issues related to hysteresis and increasing returns. However, there is an even more important reason whose effect is undeniable. The fact remains that the productive capacity of the economy itself is influenced by cumulative investment decisions based on profit expectations. Profit expectations in turn are influenced significantly by the actual output and employment path. As a result, the potential output path has no existence independent of the actual output path. Compare for instance, a 2 per cent with a 5 per cent growth of output over a run of years. The former would shape profit expectations far less optimistically geared to around 2 per cent annual growth of the market compared with the 5 per cent growth path. As a result the cumulative investment and potential output growth path associated with the 5 per cent actual growth would be considerably higher than the potential path associated with an actual 2 per cent growth. In some ways, this goes to the heart of the difference between the neo-classical way of separating the short term from the long-term economic properties of a system, and our view. We would argue that it is thoroughly misleading to try to understand the macro-dynamics of growth by attributing to the long-run output growth path an independent existence of its own irrespective of the history of the actual growth path. And, once we recognize that the rate of investment is governed by profit expectations that are shaped through the historical output growth path, an independent investment function influenced by profit expectations has to be incorporated into the analysis. It then becomes natural to separate investment from saving, and the simplistic idea of a potential output path separated from the actual output path begins to melt away in the presence of an independent investment function. Since an investment function independent of the saving function lies at the heart of the problem of effective demand, the independence of the potential from the actual output path, and the independence of investment from saving can be seen to be two inextricably linked aspects of the same problem.

Another distinguishing feature of our model has been the way it handles the stylized fact of the relative constancy of wage share in national income. Unlike in neo-classical theory, we do not view it as a technological datum. The tendency towards constancy in wage share is viewed instead as the outcome of technological competition among rival firms over market shares in the product market, as well as distributive conflict between the employers and the employees over wage claims in the labour market. Intra-class competition is interwoven with inter-class conflict to result in a race between the growth of real wage and of labour productivity. Relative constancy of wage share is seen as the outcome of this complex race with a dominant social or political dimension and, in our view has nothing to do with the Cobb–Douglas technology and the marginal productivity theory of distribution.

Technology under reasonably competitive conditions is seen to play a two-sided role in our model. On the one hand, it reduces the price level continuously by reducing production costs and helps to raise the real wage rate over time. Thus, we accept the stylized fact of rising standard of living of the working class over the longer run, and reject the classical especially Marxian postulate of a given real wage rate. On the other hand, however, the rising real wage as well as wage bargaining in the labour market put pressure on the capitalists to protect their profit by raising labour productivity. Under competitive capitalism, technological progress in this way is induced by, and also becomes a promoter of intra- and inter-class competition.

The relative constancy of the wage share is claimed to fit broadly with facts, but the various mechanisms of inter- and intra-class competition through which the wage share tends towards its constancy in this model is open to various modifications. For example, I would suggest a particularly topical one, going beyond the narrow confines of the model presented. Globalization which increases the relative importance of the external over the internal market of the national economy inevitably puts greater pressure on raising productivity for gaining international competitiveness. This mechanism of international competition has probably been quite important in driving productivity growth in the case of several countries in recent decades. While this mechanism is naturally not considered in our closed economy model, it could still help to re-emphasize one of its central results. Endogenous productivity growth driven by greater external competition would tend to leave employment growth as a residual variable in situations of mostly demand-driven growth rate of output. As a result, jobless growth, even a rising rate of unemployment becomes all the more possible in many market economies under globalization.

In the aggregative setting of our discussion, the inter-class conflict over distribution is characterized by the average productivity growth keeping pace with the average growth in the level of real wage. However, we must remember that these averages often hide more than they reveal. An increase in the mean real wage or earning rate might go hand in hand with a worsening of personal distribution of income within the wage income category. This would mean that the lower income groups or deciles lose out to the higher income deciles of wage and salary earners. As a result, while the class distribution of income remains reasonably stable with the mean wage keeping pace with productivity growth, the majority of the workers might be losing out. In such a situation the increase in mean wage would be mostly accounted for by the increase in earnings received by a small section of highly paid workers or salary earners. In a more disaggregate analysis this possibility of intra-class conflict among the workers over relative wage increases would have to be considered as a counterpart to the intra-class competition among rival capitalist firms. The intra-class competition among the workers over wage claims also tends to be accentuated by the fact that the rates of productivity growth differ among the sectors. In claiming higher wages without threatening the profit margin of firms, workers in industries marked by higher productivity growth would generally have an advantage over workers in low productivity growth sectors. The less centralized is the wage bargain, the stronger would be differentiation among the workers on this count.

A recent statistical study of productivity and real wage growth in the United States illustrates some of the issues involved:

> … over the entire period 1966–2001, as well as over 1997–2001, only top 10 percent of the income distribution enjoyed a growth rate of real wage and salary income equal to or above the average rate of economy-wide productivity growth … median real wage and salary income barely grew at all while average (mean) wage and salary income kept pace with productivity growth, because half of the income gains went to the top 10 percent of the income distribution, leaving little left over for the bottom 90 percent.
>
> (Dew-Becker and Gordon, 2005)

In such cases growing inequality in the personal distribution of income might go hand in hand with a relatively stable class distribution of income sketched out in our model.

However, we dealt with the class distribution of income at two somewhat different levels. The second lecture focused on the interaction between the class distribution of income and the level of aggregate demand. However, in the third lecture, we simplified the exposition especially with respect to the saving function, but brought into the picture the longer-run effect of distribution on endogenous productivity growth through the race between real wage and productivity growth. It would be apparent that the analysis of the model could be enriched by using more plausible savings functions, relevant especially for corporate capitalism. This might even help to trace in a parsimonious way the impact of disparate movements of wage and productivity growth in various regimes of profit- and wage-led growth in a closed economy, as well as to export – led growth in an open economy. The discussion would carry over rather easily to an open economy. If real wage does not increase in proportion to productivity, especially for workers in the export sector, the resulting decline in the unit cost of export would tend to stimulate export and aggregate demand through an export surplus. At the same time however, the higher profit share due to the slower increase in wage in relation to productivity, would also have a depressing effect on consumption. In some cases, the contraction of the internal or home market typical of a wage-led regime resulting from the relatively sluggish response of the real wage rate to productivity growth might be outweighed by the expansion of the external or foreign market. However, this scenario of export-led growth would be aborted if the contraction in home demand outweighs the stimulating effect of exports. There might be an added complication highlighted by the model in the third lecture. It pointed out that a slower increase in the real wage could also tend to reduce the pressure on firms to raise productivity over the longer run. Consequently, this particular configuration of export-led growth, driven by falling unit cost due to slower growth of wage in relation to productivity, might turn out to be somewhat problematic in the longer run. However, along the equilibrium growth path these complications are avoided by definition, because wage and productivity rise at the same rate leaving wage share and unit cost constant. Since distributional changes do not take place in the steady state, the impact of disparate movements of wage and productivity growth on investment, saving, or international price competitiveness and export can only be studied meaningfully in the out of equilibrium context.

It should be emphasized that our model, like many others, neglected one of the most crucial features of modern corporate capitalism. In understanding better the saving behaviour of a modern capitalist economy, it would certainly be important to take into account the implication of factors like the distinction between retained and distributed profit, institutional savings and distribution of income among the classes (e.g. Pitelis, 1997). Nevertheless, modification of the saving function along these lines would still leave out the important influence that the stock market or rising housing and real estate prices exert on consumption and saving which has recently attracted a lot of attention from economic commentators. The influence of stock and of the housing market on aggregate demand operates through what might be described as a 'virtual' wealth effect. This seems to me as an emerging crucial feature of modern corporate capitalism. It influences in turn many governments' stances on monetary and fiscal policy. With relatively widespread participation in the stock market within the top 10 per cent of the income receivers in the United States, a booming stock, and later a rising housing market kept consumption spending high in recent years. Thus, unlike in standard Keynesian models, despite increased inequality in the distribution of personal income, consumption expenditure was not depressed (Dayan and Maki, 2000; Maki and Palumbo, 2001). Analytically, this wealth effect is distinguishable from the more traditional 'real balance' or Pigou effect. Owners of stocks or of real estates during a boom in those asset prices become more creditworthy in the eyes of the banks and other financial institutions, and can finance higher consumption from the commercial credit system. At the same time, because banks too have a more favourable asset position, they become more willing in general to lend. It is a crucial characteristic of the virtual wealth effect that the resulting consumption boom gets financed by bank lending, and the rising debt of the asset-owning households. Indeed, it has to be financed by debt, because any attempt by the households to finance their higher consumption through realized capital gains on a large enough scale can only result in the market turning bearish, and even a crash of the stock or real estate market. By its very nature therefore, the virtual wealth effect must remain virtual, with its impact on aggregate demand and the real economy transmitted through higher lending through some system of endogenous money, with rising debt of the asset owning households as its counterpart (Bhaduri *et al.*, 2006).

Rising or falling stock prices over several periods are likely to affect investment also, but this impact is even more difficult to ascertain analytically even within the confines of the model. Following Keynes (1936), Minsky (1975, 1986) and Tobin (1969) had argued that rising stock prices make acquisition of existing real or physical asset, say of a factory, more expensive than building a new one. This substitution in favour of the cheaper option tends to increase real investment in a rising stock market. This argument holds provided the investor is an entrepreneur, set on making profit from production rather than from speculative capital gains. However, when the stock market is not dominated by entrepreneurs but by speculators, this hypothesis might not work well, insofar as rising stock prices fuel further expectations of capital gains. Indeed the lure of substansial capital gains might persuade even an entrepreneur to turn into a speculator!

Leaving aside these complications, it can be easily seen that any significant positive impact of the stock market on investment expenditure would reinforce the profit- or investment-led regime discussed earlier in Lecture II in mutually reinforcing ways. In a closed economy, a positive virtual wealth effect through the operation of the stock market would raise consumption expenditure, while it would also stimulate investment. This would typically be a path led by expanding consumption and investment expansion, but with rising indebtedness of households, and increasing leverage ratio for the firms.

However, it need not necessarily be interest-bearing debt for the firms. With rising stock prices, firms can usually raise the same amount of finance more comfortably from the market with fewer new issues, and less fear of hostile take-over. This might act as a more important stimulant to investment than the Keynes–Tobin–Minsky two-price arbitrage view mentioned earlier. If firms take recourse to issuing new equities instead of relying on commercial borrowing to raise the required finance for investment, investment-led expansion becomes equity- rather than debt-driven, although households might continue to increase their indebtedness with the commercial banks in a bullish stock market (Taylor, 2004). In a very simple comparative static exercise conducted in Lecture II, we analysed how higher stock price, through its effects on investment and saving might affect capacity utilization and income distribution in situations of forced saving and profit squeeze.

Our analysis throughout has been based on the assumption that productivity growth takes place in the shadow of innovations, but the two are not the same. A wedge between them is driven by the need for diffusion which depends in turn on the market structure and intellectual property rights regime. It was pointed out that neo-classical growth theory fails to distinguish sufficiently between the potential and the actual output path. In a similar way, it also overlooks the distinction between potential and actual labour productivity growth due to innovation. Innovation is defined broadly as improvement in productive knowledge. Clearly, it raises the potential but not necessarily the actual productivity of labour. And yet, this distinction is hard to maintain within a framework that emphasizes exclusively the supply side. Many technologically viable innovations promising potential for increase in productivity may not be realized, either because of inadequate demand, or because of the nature of competition or, for other reasons like imperfect capital market that make their adoption difficult or too risky and premature in given circumstances (Diamond, 1999). Two comments are in order in this context. First, in many empirical works, the number of research papers or patents often serves as a proxy for innovation. However, they relate to the potential, but not necessarily the actual increase in productivity creating a possible disjunction between innovation and productivity growth. Second, the growth in actual productivity can even exceed at times the growth in potential productivity due to the existence of unused patents, operating as an unused stock of inventories of knowledge.

Adam Smith's celebrated observation that the division of labour is limited by the extent of the market pointed to the inhibiting role of a limited market size on productivity growth. Some potentially productivity-raising innovations might not get adopted due to limited market size, with the result that the potential would not be translated into actual labour productivity growth. The opposite might also happen. In a rapidly expanding market, the inventory of unused knowledge or patents might be used up at a faster rate, with the result that actual productivity growth would exceed potential productivity growth, at least for a while. There is a distinct possibility of reinforcing positive feedbacks and cumulative causation here. For instance, a sluggish growth in the real wage rate reduces the pressure on firms to adopt labour productivity raising innovation. At the same time, in a wage-led regime slow growth in real wage also means a slowly expanding size of the market. These two forces would tend to reinforce one another in slowing down the growth rate.

Not only the possible inhibiting role of a limited of a market size, but also the negative effect of competition on the adoption or diffusion of innovation goes almost unnoticed in the mainstream literature. A most conspicuous example of the inhibiting influence of competition in real life is 'preventive' or 'pre-emptive patents'. A significant number of innovations are regularly bought up by large corporations from private innovators in universities and research institutes with a view to preventing their possible use by rivals. In these cases, patent rights are turned almost into their opposite. Patent rights retard rather than generate new knowledge. The individual innovator might make some money by selling off his or her innovation, but the society gains nothing from it, and market competition helps to 'kill' the innovations. Similarly competition leads to trivial patent differentiation, i.e. almost the same product trivially differentiated is patented as a new product to charge a higher price to the consumer.[13] At times, patent rights also become an instrument in the hand of large corporations for harassing independent individual researchers with law suits. Although hardly mentioned in standard literature, phenomena like preventive patents and trivial patent differentiation show how intellectual property rights can have two entirely different faces. They not only promote but at times also demote innovations to illustrate the complex relation that exists between the knowledge-generating sector and the knowledge-using sector, especially under corporate capitalism. It is perhaps not too much of an exaggeration to say that only when a corporation does in-house research, can we broadly assume that the productive knowledge that it generates would be put to use. However, with a network spread across universities, independent and publicly funded research institutes, a floating body of consultants, and in-house research for profit in large corporations, the link between potential and actual productivity growth becomes considerably more complex for any given regime of intellectual property rights (Gilsing, 2005).

Standard literature does recognize to some extent that competition in the knowledge-using product market can have either a positive or a negative effect on innovations in the knowledge-generating sector (e.g. Aghion and Howitt, 1998: Ch. 7). In general, the smaller the size of the firm, the weaker would be both its risk-taking ability, and its budget for research. The result would be an inverse relation between the number of firms, and the potential productivity growth through innovations that are generated by the firms. On the other hand, if we assume as a rule of thumb that the number of firms serves as an index of the intensity of competition in an industry, then in a given regime of intellectual property rights, actual productivity growth would tend to be higher through faster diffusion of technology due to greater competition, as the number of firms increases.[14] This is shown in a highly simplified manner in the following Diagram 8.

In Diagram 8 the PP line represents the potential productivity growth line. It is negatively sloped on the assumption that the larger the number of firms the smaller is their size, and the lower is the capacity of an individual firm either to devote adequate budget or bear the risk of trying to innovate.

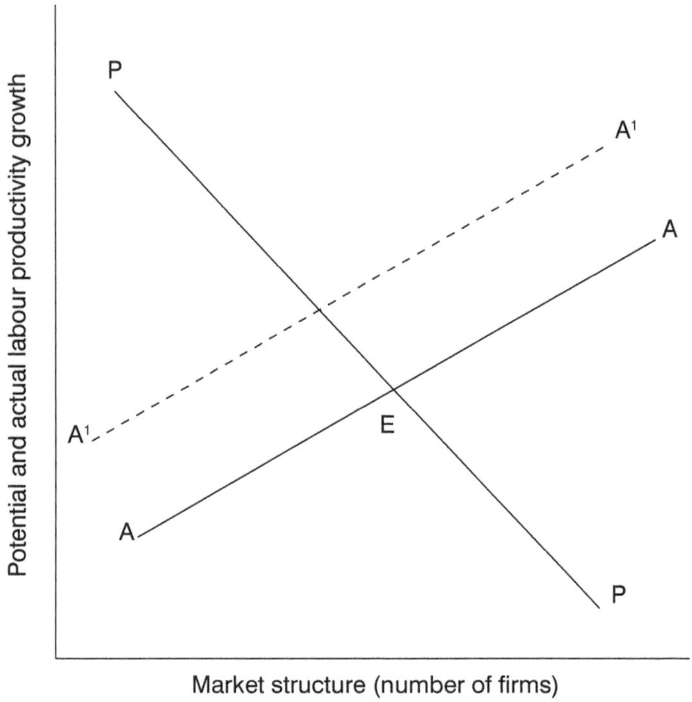

Diagram 8

From this point of view, competition among many small firms tends to discourage the rate at which new productive knowledge is generated. On the other hand, however, given the regime of intellectual property rights, the larger the number of firms the greater is the scope of mostly imitative diffusion of technology at little cost. This faster pace of diffusion in a more competitive market structure tends to make actual productivity growth, represented by the AA line, an increasing function of the number of firms. The intersection between the potential and the actual productivity growth curve at E determines the observed productivity growth where the flow of new productive knowledge is being fully utilized.

With this diagram, we could also perform a few simple thought experiments, similar to standard comparative static exercises, but without the stability considerations. A more accommodative intellectual property rights regime would further speed up the rate of technological diffusion, and shift the AA curve upwards to A^1A^1, represented by the broken line, and the extent of the shift would tend to be smaller in general, the costlier it is to adopt the new technology.

This simple diagram could be treated as a convenient device to sort out some of our ideas regarding the interrelation between market structure, intellectual property rights and potential and actual productivity growth. The firms would usually operate within the constraint set by the lower envelope AEP. However, insofar as unutilized stock of past innovations might be available for exploitation, the actual growth rate of productivity may even exceed the potential growth rate for some time due to the pressure of competition, shown by the region to the right of the point E. In the absence of unused inventories of patented knowledge, the locus of the points of intersection E traces out the observed productivity growth over time through the interaction between knowledge generation and its diffusion. If we take the view, similar to that of Schumpeter particularly in his early writing (1961) that the generation of innovation is governed by the possibility of abnormal profit which is transient due to diffusion, then the shifts in the potential (PP) and the actual (AA) productivity growth curves cannot be treated as entirely independent. For instance, a less strict property rights regime shifts the AA curve upwards to A^1A^1, but would also shift simultaneously downwards the PP curve due to faster dissipation of abnormal profits through speedier diffusion.

A point of caution is needed before this utterly simple diagram is taken too literally. While considering shifts of the AA or PP curves, the degree of competition measured by the number of firms on the horizontal axis needs to be interpreted carefully. They represent hypothetical rather than real change in competition. This means that for a given shift representing some relaxation in intellectual property rights for example, not only is productivity increased along the vertical axis. On the horizontal axis it becomes in effect equivalent to a change in the degree of competition, measured by a corresponding hypothetical change in the number of firms.

Since the locus traced out by the intersection point E of shifting PP and AA curves through time might be positively or negatively sloped, little can be said in general about the relation between competition (measured by the number of firms) and the observed productivity growth in an economy. The central justification of patent and intellectual property rights is the idea that firms are induced to innovate by the lure of temporary monopoly rent. This implies that faster diffusion of technology would also compete away this rent quicker in a more intensely competitive product market. From this point of view, the effect of competition on potential productivity growth would tend to be negative, as shown by the PP curve. Empirical evidence on the other hand seems to suggest a positive correlation between a observed productivity growth and some index of competition like the number of firms in an industry or the inverse of market share (Blundell *et al.*, 1995; Nickell, 1996). This empirical result could be interpreted in terms of Diagram 8 to mean that the locus traced out by the point E has a positive slope over time through the interaction between potential and actual productivity growth.

The impact of a higher rate of growth of demand on productivity growth could also be exhibited in the same diagram through an upward shift of the AA curve. This would be in conformity with the Kaldor–Verdoorn law, which postulates that higher output growth, as a proxy for the rate of expansion of the size of the market, affects positively productivity growth. Similarly, a parametric shift in technology policy, e.g. devotion of more resources for research and development, could be captured by an upward shift of the potential productivity curve PP. In short, this simple diagram is designed for a preliminary qualitative sorting out of ideas about the various factors that come into play in determining observed productivity growth. However, the main reason for constructing this diagram is more important. It exhibits how productivity growth takes place as a result of the interaction between potential and actual productivity growth in different climates of competition and market structure, and responds to various shift parameters like growth in demand, change in intellectual property rights, resource devoted towards research and development etc. In this process, both the 'supply side' of potential productivity growth, and the 'demand side' of actual productivity growth play their roles.

It is indisputable that learning by the various agents involved in generating and using knowledge play a critical role in productivity growth (David, 1975). Therefore it would be useful to try to incorporate the idea of learning in this framework of interaction between the growth in potential and actual productivity. With this in view, we begin from the distinction made by Rosenberg (1982; 121–2) between three components of learning: (a) learning in the R&D stage or learning by searching; (b) learning at the manufacturing stage or learning by doing, mostly for cutting down production cost of designing a technology or a method of production; and finally, (c) learning through operating the technology, by gaining skill and expertise in using the given production method through experience (see also Mulder, 2005). Following this threefold classification, we might roughly say that the first type of learning would shift potential productivity curve PP upwards over time. The second type of learning might shift upward the potential productivity curve in some cases; but it would also shift the actual productivity curve upwards insofar as the more cost-effective designs would make its adoption easier for the firms in any given market structure. And, the third type of learning would shift the actual productivity curve AA continuously upwards through time. As a result of these different components of learning, both the curves PP and AA would tend to drift upwards through time at different speeds, and trace out a locus of observed equilibrium points E over time with a positive or negative slope, depending on how these different components of the learning rate play out through time.

A particularly unsatisfactory aspect of our formal analysis in the model has been its implicit assumption that individual firms tend to innovate in isolation. This is a common assumption, frequently made by model builders. And yet, it seems increasingly less tenable under corporate capitalism. As a matter of fact, firms often depend on other firms or research centres in the real world for complementary knowledge, and tend to cooperate through either formal arrangements like licensing fees, or informal ones like inter-firm mobility of consultants and a host of similar devices (Baumol, 2004).[15] Insofar as one normally expects considerations of strategic interaction to be usually more compelling in oligopolistic market structures dominated by a few large firms, some degree of complementarity in research through mutually beneficial cooperation becomes quite common, especially in the market for knowledge generation than in the market for products for using those knowledge. However, networking and complementarity in research among private firms are almost invariably intertwined with competition especially over shares in the product market. Therefore, the trajectories of innovations tend to be driven by a complex process of technological cooperation as well as competition among a few large firms in a market economy. Corporations, if not the atomistic firms of the textbook variety, are caught in situations of strategic interaction. They must take into account, even anticipate, the reactions and moves made by other firms. In that context both cooperation and competition become useful strategies depending on the particular situation. For this reason is considerably easier to model technological interaction among many firms, as we tried in our model, but extremely difficult to construct any sufficiently general plausible model of competition among a few large firms.

The focus of our formal model has been on the diffusion of technology with numerous firms. Diffusion clearly affects actual rather than potential productivity growth. Nevertheless, in one sense some of the conclusions of this model focusing on diffusion might be less restrictive than appears at first sight, because, in the very process of diffusion, adaptations are made, and often new, even dramatic innovations come from that source, a bit like mutation of the copying gene (Diamond, 1999). It is more difficult to capture how the knowledge for innovation is generated than to try to understand the diffusion process with adaptation. At least as a research strategy at this stage, it seems more manageable, and probably more worthwhile to focus on the diffusion of technology in relation to market structures. In this sense, the approach of this lecture has also been different from the current crop of endogenous growth models. They are forced to make implausible heroic assumptions about various knowledge-generating production functions without making a clear distinction between the potential and the actual growth of productivity.

As already said, oligopolistic market structure is more difficult to theorize about, because many equally plausible models of strategic interactions can be constructed, depending on the particular rules the game set for the interaction to take place. It deserves to be emphasized that these lectures do not underplay the role of interaction, but address this problem of interaction among agents from a very different angle. Strategic interaction, as visualized in game theory, is based on conscious and rational choice of strategies by players, taking into account the possible strategies or reactions of other players. Therefore, by its very nature, game theory deals with situations involving two or more agents. So does Keynesian macroeconomics, but in a different way, for example, by making a distinction between decisions of the savers as one agent, and of the investors as another agent, and perhaps the banks as yet another agent which decides to finance investment projects. It is an institutional feature of capitalism that the decision to invest is taken by firms mostly on expectations of future profits, and most of the investment is financed by financial intermediaries like the banks, while saving comes from the income the households receive. To emphasize this institutional aspect, Hicks (1974) had aptly characterized this aspect of the capitalist economy as an 'overdraft' economy. It is the structural feature of an overdraft economy that separates investors from savers, giving investors the autonomy to decide on the level of investment through bank loans irrespective of the decision of the savers. In this way, the interactions postulated in Keynesian macroeconomics become structural rather than strategic. The interaction between the investors and the savers as two agents, generate the familiar circular flow of income, the problem of possible deficiency of demand etc. They result, for example, predominantly in income adjustment in a particular economic situation, while in another situation these structural interactions give rise to inflation and distributional changes. Neither income adjustment nor distributional changes are the consequences of strategic choices by individual agents; instead, they are to be treated as outcomes of the logic of the particular situation, e.g. existence of excess capacity in one case, or strong collective bargaining institutions in the other case. This is the fundamental difference between macroeconomic theory based on structured or institutionalized interactions among agents, and microeconomic theory based on strategic choices by rational players in a game-theoretic set-up. Perhaps the closest analogy explaining this type of structural interaction is the biological interaction between two interacting species in a given environment. Blurring this distinction between structural and strategic interaction, and trying to reduce every aspect of economic theory to choice of strategies by rational individuals would not be healthy either for macro- or for micro-economic theory.

Notes

1 In the presence of technical progress the labour force in natural units becomes more productive in efficiency units, e.g. when one worker does the work of two, the number of natural labour is multiplied by a factor of two. In general, labour in efficiency units $L_E = Q.L$, where Q in the multiplicative factor on labour, L in natural units. In terms of growth rates (i.e. by logarithmic differentiation with respect to time) this means, for example that a 2 per cent growth in labour productivity along with a 1 per cent growth in the natural labour force equals a 3 per cent growth of labour in efficiency units.

2 On the assumption of a constant proportional mark-up on unit variable cost assumed to be labour cost only in a closed economy, profit share (h) and mark-up (m) are positively related by equation (2) in the text.

3 This acts like an 'inflationary barrier' (Robinson, 1956). Therefore, pure quantity adjustment in this case might co-exist with inflation characterized by both the nominal price level and the money wage rate rising at the same rate.

4 The local stability of (8) and (9) requires the trace of the relevant Jacobian evaluated at equilibrium to be negative, which is equivalent to condition (13). However, the determinant of the Jacobian is zero because the two equations are linearly dependent. This implies a continuum of equilibria along the *IS* curve, while the stability of any equilibrium point on the *IS* curve requires that the only non-zero characteristic root is negative, i.e. the trace condition.

5 Harrod neutrality of technical progress is defined as constant capital output ratio at a constant rate of profit. This implies constancy of wage share, and is ensured if the real wage rate rises at the same rate as labour productivity. Interpreted in terms of the aggregate production function with usual properties, this means that the natural labour force is augmented in efficiency units due to Harrod-neutral technical progress. See also note 1.

6 If ρ is the expected and r the actual rate of profit, then $g = f(\rho)$, and $\rho = \varphi(r)$, $\phi'(r) > 0$. On substitution, g becomes an increasing function of r, as in equation (20) in the text. This otherwise trivial extension is of some interest, when the expected rate of profit reacts strongly to the actual profit rate in a highly optimistic economic environment, and the elasticity of expectation is greater than unity to destabilize the system.

7 In Nash equilibrium, players accept the equilibrium because it is within nobody's power to improve upon the situation. Such a construction of equilibrium in the

present context would be misleading. For instance, the workers' perception of risk and urgency concerning the need to change an equilibrium might alter with a rising rate of unemployment. For instance, with 5 per cent unemployment and reasonable social security, workers might accept that it is not within their power to change the ongoing state of affairs. They would not risk changing the equilibrium. However, with say 25 per cent unemployment and worsened social security, they might become desperate, and test their power to risk changing the equilibrium. Unlike a well-defined strategy set, the perceived limits to power of a class or even of an individual is ill defined, and not context-free.

8 See Marglin (1984) and Pitelis (1997) for a discussion of the saving function under corporate capitalism.

9 On the assumption that markets are always cleared, it has become customary to treat saddle-path solutions as depicting stable configurations with suitably defined 'jump variables' which place the system on the stable arm of the saddle-path (see Gondolfo, 1996: Ch. 22 for a summary). However, unless it is shown how a decentralized market economy places itself invariably on the stable arm of the saddle-path, this 'new classical' assumption remains an article of faith of little relevance in the present context.

10 The innovating atomistic firm is assumed to engage a marginal amount of labour at higher productivity, while the rest of the labour force is engaged by numerous other firms in the rest of the economy to define the average productivity. Because the innovating firm is atomistically small, the higher productivity of labour engaged by it has a negligible quantitative effect on the average labour productivity of the economy.

11 This assumption of 'atomistic' firm as innovator, considered typically by the classical economists, corresponds to a picture where technology is generated by isolated firms or individuals, and its diffusion occurs gradually among numerous small firms through competition for market shares. This requires the assumption of price setting behaviour by the innovating firm even under competitive conditions, because it perceives some demand constraint, and therefore tries to increase its market share by lowering price. Needless to add, this classical view of competition is incompatible in many ways with the neo-classical view of perfect competition, where all firms are price-takers, perceiving no demand constraint at that price. It must also be added that the process of diffusion is likely to be very different in oligopolistic market structures (see Lecture IV)

12 This also seems more compatible with the long-term historical trends in real wage and productivity growth in most industrial economies.

13 According to a recent study, patented drugs are 300–400 per cent more expensive than non-patented drugs, and about 210 billion dollars are spent on patented drugs in the US. And yet, according to the Federal Food and Drug Administration Authority, some 70 per cent of patented drugs are trivially differentiated (Baker, 2005).

14 On the assumption that either the capital market is perfect, or the cost of adopting new innovations are not prohibitively high even for the small firms.

15 International patent filings under the Patent Cooperation Treaty in 2003 showed that the US has a large lead (39,250) followed by Japan (16,774), Germany (13,979), the UK (6,090), France (4,273) and the Netherlands (4,180). The networking of research also seems more pronounced in the US. It is shown partly by the fact that universities like California (438 patents), Caltech (139) and MIT (127) contribute significantly to this list (Harding, 2004: 52).

References

Aghion, P. and Howitt, P. (1998) *Endogenous Growth Theory*, ch 7: "Market Structure", Cambridge, MA and London: MIT Press, pp. 205–32.

Arrow, K. (1962) 'The Economic Implications of learning by doing', *Review of Economic Studies*, 29: 55–73.

Arthur, W.B. (1994) *Increasing Returns and Path Dependence in the Economy*, Ann Arbor, MI: Michigan University Press.

Baker, D. (2005) 'The reform of intellectual property', *Post-Autistic Economic Review*, issue 32 (http://www.paecon. Net/PAERReview/ issue 32/Baker32.htm.

Barro, R. and Sala-i-Martin, X. (1995) *Economic Growth*, New York: McGraw-Hill.

Baumol, W.J. (2004) *The Free-Market Innovation Machine: Analysing the Growth Miracle of Capitalism*, New Delhi: Oxford University Press.

Bhaduri, A. (1983) 'Multimarket classification of unemployment: a skeptical note', *Cambridge Journal of Economics*, 7: 235–41.

Bhaduri, A. (2002) 'Chaotic implications of the natural rate of unemployment', *Structural Change and Economic Dynamics*, 3(3): 357–65.

Bhaduri, A. (2006) 'Endogenous economic growth: a new approach', *Cambridge Journal of Economics*, 30: 69–83.

Bhaduri, A. (2006) 'On the dynamics of profit and wage led growth', *Cambridge Journal of Economics*, forthcoming.

Bhaduri, A. and Harris, D.J. (1987) 'The complex dynamics of the simple Ricardian system', *Quarterly Journal of Economics*, 102: 893–902.

Bhaduri, A. and Marglin, S. (1990) 'Unemployment and real wage: the economic basis for contesting political ideologies', *Cambridge Journal of Economics*, 14: 375–93.

Bhaduri, A., Laski, K. and Riese, M. (2006) 'A model of interaction between the real and virtual economy', *Metroeconomica*, 57: 412–27.

Blackburn, K. (1999) 'Can stabilisation policy reduce long-run growth?', *Economic Journal*, 109: 67–77.

Blanchflower, D.G. and Oswalid, A.J. (1994) *The Wage Curve*, Cambridge, MA: MIT Press.

Blanchflower, D.G. and Oswalid, A.J. (1995) 'An introduction to the wage curve', *Journal of Economic Perspective*, 9: 153–67.

Blundell, R., Griffith, R. and van Reenen, J. (1995) 'Dynamic count data model of technological innovation', *Economic Journal*, 105(429): 333–44.

Card, D. (1995) 'The wage curve', *Journal of Economic Literature*, 33:785–99.

Cass, D. (1965) 'Optimum growth in an aggregative model of capital accumulation', *Review of Economic Studies*, 32: 233–40.

Chiarella, C. and Flaschel, P. (2000) *The Dynamics of Keynesian Monetary Growth: Macrofoundations*, Cambridge: Cambridge University Press.

David, P. (1975) *Technical Change, Innovation and Economic Growth*, Cambridge: Cambridge University Press.

Dayan, K.E. and Maki, D.M. (2000) 'Does stock market wealth matter for consumption?', Mimeo. Board of Governors of the Federal Reserve System.

Dew-Becker, I. and Gordon, R.J. (2005) 'Where did the productivity growth go? Inflation dynamics and distribution of income', *National Bureau of Economic Research*, Working Paper 11842 (http://www.nber.org/papers/w11842).

Diamond, J. (1999) *Guns, Germs and Steel*, New York: Norton, pp. 239–64.

Diamond, P. (1965) 'National debt in a neoclassical growth model', *American Economic Review*, 52: 1126–50.

Foley, D.K. and Michl, T.R. (1999) *Growth and Distribution*, Cambridge, MA: Harvard University Press, especially Chapters 5, 10, 13.

Frankel, M. (1962) 'The production function in allocation and growth model', *American Economic Review*, 52: 995–1002.

Friedman, M. (1968) 'The role of monetary policy', *American Economic Review*, 58: 1–17.

Gandolfo, G. (1996) *Economic Dynamics* (3rd edn), Berlin: Springer, Chapter 23.

Gilsing, V. (2005) *The Dynamics of Innovation and Interfirm Networks: Exploration, Exploitation and Co-evolution*, Aldershot: Edward Elgar,

Harding, A. (2004) *The Scientist*, 27 September, p. 52.

Harrod, R.F. (1939) 'An essay in dynamic theory', *Economic Journal*, 49: 14–33.

Harrod, R.F. (1942) *Towards a Dynamic Economics*, London: Macmillan.

Hicks, J.R. (1965) *Capital and Growth*, Oxford: Clarendon Press

Hicks, J.R. (1974) *The Crisis of Keynesian Economics*, Oxford: Basil Blackwell.

Kahn, R.F. (1959) 'Exercise in the analysis of growth', *Oxford Economic Papers*, 11: 143–56.

Kaldor, N. (1956) 'Alternative theories of distribution', *Review of Economic Studies*, 23: 83–100.

Kaldor, N. (1957) 'A model of economic growth', *Economic Journal*, 67: 591–624.

Kaldor, N. (1989) 'The role of increasing returns, technical progress and cumulative causation in the theory of international trade and economic growth', in F. Targetti and A.P. Thirlwall (eds) *The Essential Kaldor*, London: Duckworth, pp. 327–51.

Kalecki, M. (1971) *Selected Essays on the Dynamics of the Capitalist Economy*, Cambridge: Cambridge University Press.

Keynes, J.M. (1930) *A Treatise on Money, Volume 1*, London: Macmillan.

Keynes, J.M. (1936) *The General Theory of Employment, Interest and Money*, London: Macmillan.

Keynes, J.M. (1937) 'The general theory: fundamental concepts and ideas', *Quarterly Journal of Economics*, 51: 1–15.

Koopmans, T.C. (1965) *On the Concept of Optimal Economic Growth, in the Econometric Approach to Development Planning*, Amsterdam: North Holland (for Pontificia Academy).

LaSalle, J.P. and Lefschetz, S. (1961) *Stability of Liapunov's Direct Method with Applications*, New York: Academic Press.

Leijonhufvud, A. (1968) *Keynesian Economics and the Economics of Keynes*, London: Oxford University Press.

Lucas, R. (1988) 'On the mechanics of economic development', *Journal of Monetary Economics*, 22: 3–42.

Maki, D.M. and Palumbo, M.G. (2001) 'Disentangling the wealth effect: a cohort analysis', Mimeo, Board of Governors of the Federal Reserve System.

Malinvaud, E. (1977) *The Theory of Unemployment Reconsidered*, New York: John Wiley and Sons.

Marglin, S.A. (1984) *Growth Distribution*, Cambridge, MA: Harvard University Press. Also, Pitelis (1997).

Marshall, A. (1920) *Principles of Economics* (8th edn), London: Macmillan.

Marx, K. ([1867] 1954) *Capital*, Vol. 1. Moscow: Progress Publishers; English translation of *Das Kapital*, Vol. 1, 1867, Hamburg: Meissner.

Minsky, H.P. (1975) *John Maynard Keynes*, New York: Columbia University Press.

Minsky, H.P. (1986) *Stabilizing an Unstable Economy*, New Haven, CT: Yale University Press.

Mulder, P. (2005) *The Economics of Technology Diffusion and Energy Efficiency*, Aldershot: Edward Elgar.

Nickel, S.J. (1996) 'Competition and corporate performance', *Journal of Political Economy*, 104(4): 724–46.

Pasinetti, L.L. (1959–60) 'A mathematical formulation of the Ricardian system', *Review of Economic Studies*, 26: 78–98.

Pasinetti, L.L. (1962) 'The rate of profit and income distribution in relation to the rate of economic growth', *Review of Economic Studies*, 29: 267–79.

Pasinetti, L.L. (2000) 'Critique of neoclassical theory of growth and distribution', *Banca de Lavoro Quarterly Review*, 53: 383–432.

Petty, W. (original 1662) 'A treatise on taxes and contribution', in C.H. Hull (ed.) *The Economic Writings of Sir William Petty*, Cambridge: Cambridge University Press, 1899; revised Augustins M. Kelly edition, New York, 1963.

Pitelis, C. (1997) 'On Kaldor and pensions', *Cambridge Journal of Economics*, 21: 469–82.

Ramsey, F. (1928) 'A mathematical theory of saving', *Economic Journal*, 38: 543–99.

Ricardo, D. (original 1817) 'Principles of political economy and taxation', in P. Sraffa (ed.) *Works and Correspondence of David Ricardo*, Cambridge: Cambridge University Press, 1951, vol. 1.

Robinson, J. (1956) *The Accumulation of Capital*, London: Macmillan.

Robinson, J. (1962) 'A simple model of accumulation in her Essays in the Theory of Economic Growth', London: Macmillan.

Robinson, J. (1964) 'Pre-Keynesian theory after Keynes', *Australian Economic Papers*, 3: 25–35.

Romer, D. (1996) *Advanced Macroeconomics*, New York: McGraw-Hill.

Romer, P. (1986) 'Increasing return and long-run growth', *Journal of Political Economy*, 94(5) (October): 1002–37.

Romer, P. (1990) 'Endogenous technical change', *Journal of Political Economy*, 98(5, part 2) (October): S71–S102.

Rosenberg, N. (1982) *Inside the Black Box: Technology and Economics*, Cambridge: Cambridge University Press.

Samuelson, P.A. (1958) 'An exact consumption-loan model of interest, with or without the social contrivance of money', *Journal of Political Economy*, 66: 467–82.

Samuelson, P.A. (1966) 'A summing up', *Quarterly Journal of Economics*, 80: 568–83.

Schumpeter, J.A. (1939) *Business Cycles*, New York: McGraw-Hill.

Schumpeter, J.A. (1942) *Capitalism, Socialism and Democracy*, New York: Harper and Brothers.

Schumpeter, J.A. (1961) *Theory of Economic Development*, New York: Oxford University Press (original, 1911; English translation, 1934).

Smith, A. (1776) 'An inquiry into the nature and causes of the wealth of nations, the Glasgow edition of the works and Correspondence of Adam Smith' (vol 2), in R.H. Campbell, A.S. Skinner and W.B. Todd (eds), Oxford: Oxford University Press.

Solow, R. (1956) 'A contribution to the theory of economic growth', *Quarterly Journal of Economics*, 70: 65–94.

Solow, R. (1957) 'Technical change and the aggregate production function', *Review of Economics and Statistics*, 39(August): 312–20.

Solow, R. (2000) 'Lessons of suggestion for aggregative growth theory' in *Growth Theory*, An Exposition (2nd edn), Ch. 12, New York: Oxford University Press, pp. 180–6.

Sraffa, P. (1960) *Production of Commodities by means of Commodities: Prelude to a Critique of Economic Theory*, Cambridge: Cambridge University Press.

Steindl, J. (1952) *Maturity and Stagnation in American Capitalism* (2nd edn), New York: Monthly Review Press.

Swan, T. (1956) 'Economic growth and capital accumulation', *Economic Record*, 32: 343–61.

Taylor, L. (1983) *Structural Macroeconomics*, New York: Basic Books.

Taylor, L. (2004) *Reconstructing Macroeconomics*, Cambridge, MA: Harvard University Press, Chapters 4, 5, 7.

Tobin, J. (1969) 'A general equilibrium approach to monetary theory', *Journal of Money, Credit and Banking*, 1: 15–29.

Uzawa, H. (1961) 'Neutral inventions and the stability of growth equilibrium', *Review of Economic Studies*, 28.

Young, A. (1928) 'Increasing returns and economic progress', *Economic Journal*, 38: 527–42.

Index

www.ingramcontent.com/pod-product-compliance
Ingram Content Group UK Ltd.
Pitfield, Milton Keynes, MK11 3LW, UK
UKHW022311250926
475702UK00046B/675
* 9 7 8 0 4 1 5 4 2 1 0 8 9 *